Oxford University Committee for Archaeology
Monograph 11

Archaeological Results
from
Accelerator Dating

Research contributions drawing on radiocarbon dates
produced by the Oxford Radiocarbon Accelerator
based on papers presented at the SERC sponsored conference
'Results and Prospects of Accelerator Dating'
held in Oxford on October 1985

edited by
J. A. J. Gowlett and R. E. M. Hedges

Oxford University Committee for Archaeology
1986

Published by
Oxford University Committee for Archaeology
Institute of Archaeology
Beaumont Street
Oxford

Distributed by
Oxbow Books
10 St. Cross Road, Oxford OX1 3TU

© J. A. J. Gowlett, R. E. M. Hedges and individual contributors, 1986

ISBN 0 947816 11 9

Typeset at Oxford University Computing Service
Printed in Great Britain
at the Alden Press
Oxford

EDITORS' PREFACE

J. A. J. Gowlett and R. E. M. Hedges

Radiocarbon dating by accelerator mass spectrometry (AMS) has been taking place fairly routinely since 1983. This phase of operational dating came after a prolonged period of initial development, and it is perhaps surprising how rapidly the technique has advanced towards maturity in only three years. Enough dates have now been produced, concentrated on selected archaeological topics of major interest, to allow accelerator dating to be evaluated in practice on a broad basis. To this end a conference was organised in Oxford in October 1985 under the auspices of SERC. Here 'consumers' of dates were able to present case studies of dating by AMS, and to assess the results in their archaeological context.

This volume results from that conference; it also incorporates additional material, including some results of dating which has taken place only in the last few months, together with one or two other papers which are also relevant to the problems which face archaeologists and dating scientists in their dealings with one another.

The Research Laboratory for Archaeology always intended that the Oxford accelerator should concentrate on archaeological problems of broad importance, so that coherent patterns of dates would be achieved, and the power of the technique would not be dissipated. Professor Renfrew and Professor Jope, as members of the SERC steering panel, both argued strongly for this approach in the early days of programme formulation, when even basic requirements for technical success were still some distance away.

Now several hundred dates have been produced, both in the laboratory's in-house programme, and in the National Facility programme, which has been carried out according to the recommendations of a programme advisory panel appointed by SERC. Our contributors in those programmes have provided material on some of the most interesting problems in prehistory, on a thematic basis. In many of the examples very significant progress has been made in investigating sites or subjects which were previously out of reach of radiocarbon. The dating work is still in progress on most of these themes, and there is no doubt that much more will be understood with a great deal more certainty when fifty or a hundred, rather than ten or twenty, dates have been achieved on some of these topics.

Interpretation is therefore at a provisional stage. But although that makes this book an interim statement of progress, it is probably also fair to say that it marks a stage in the application of AMS to archaeology where we have come to the end of the beginning. We cannot even glimpse how far the process will develop, though further technical improvements are envisaged, and seem bound to ensure the throughput of much larger numbers of dates, and probably an extension of the radiocarbon time range (see Hedges, this volume).

We have impinged upon the contributors to an absolute minimum in the editorial sense, partly so as to produce this volume as quickly as possible, and partly so as to leave well alone. On occasion, a laboratory as 'producer' will take a slightly different view of problems from the archaeologist as 'consumer'. Our approach in the laboratory has been to vet the output of dates, to try to pin down 'unacceptable' or 'rogue' dates as early as possible, and to redate material where questions remain. Naturally enough some problems are unresolved: but much of the interest lies in these.

The main difficulty in achieving uniformity has been that of terminology. We ourselves stick steadfastly to the radiocarbon convention, reaffirmed recently at the Trondheim Conference, of referring to uncalibrated dates on the BP scale. That conference decision must have the inevitable effect that the 'bp/BP' convention will soon disappear as calibrated dates are referred to as 'Cal BP' (see Gillespie and Gowlett, this volume). We have not however attempted to dissuade contributors from the use of ad/bc for uncalibrated dates, since it seems to provide the easiest way of maintaining continuity with discussions in older literature. Nevertheless this convention will tend to disappear as new universally agreed calibration curves are introduced.

ACKNOWLEDGEMENTS

SERC gave support for the conference. We thank especially Evelyn Hendy for organising the compilation of typescripts onto computer files, and David Brown and Val Tomlin of Oxbow Books for the subsequent production.

CONTENTS

IV. Later Prehistory

V. Radiocarbon Methodology and Technology

FOREWORD

D. R. Harris
(Chairman, Programme Advisory Panel)

As the Editors point out in their Preface, it was envisaged from the founding of the Oxford Radiocarbon Accelerator Unit that dating should focus on themes and problems of major archaeological significance. The original aims of the Steering Panel set up by the Science and Engineering Research Council (SERC) to guide the work of the Unit included the *initiation* of archaeological dating projects, but this was seen to be incompatible with SERC practice (for research-grant applications) which requires the initiative to be taken by individual scientists in British universities and polytechnics, and the applications to be subject to peer review. In 1983 the present Programme Advisory Panel was established (as a sub-committee of the SERC Science-based Archaeology Committee) to monitor the work of the Unit and to screen applications for accelerator dating under the National Facility programme, which occupies half of the laboratory's capacity.

Since then the Panel has evolved a policy by which all applications for National Facility dates are judged independently on their scientific merit, and at the same time several major archaeological themes have been designated as especially appropriate for accelerator dating. The choice of themes has been partly guided by the dating experience that the Unit gained during its first five years, and the themes themselves are kept under review by the Panel. At present three such themes are emphasised in the National Facility programme: plant and animal domestication and the origins of agriculture; the Upper Palaeolithic in Europe and South-west Asia; and later British prehistory, with emphasis on the dating of Neolithic burials and Bronze Age metalwork. Other themes have been pursued by the Unit in its in-house programme, notably the controversial question of early man in America, to which a major contribution has been made (Gillespie *et al.* 1985, pp. 240–241; Taylor *et al.* 1985).

At the conference held in October 1985, from which this book arises, the contribution so far made by accelerator dating at Oxford to the three main National Facility themes was reported and discussed. Most of the papers given related to the three themes, and at the conference it became clear that impressive advances had already been made in resolving chronological problems in each of these areas of archaeological enquiry. Some of the dating achievements and prospects relating to the first of these areas — plant and animal domestication and the origins of agriculture — are reviewed in the next section.

REFERENCES

Gillespie, R., Gowlett, J.A.J., Hall, E.T., Hedges, R.E.M. and Perry, C., 1985, Radiocarbon dates from the Oxford AMS system: Archaeometry Datelist 2, *Archaeometry* 27, 2, 237–246.
Taylor, R.E., Payen, L.A., Prior, C.A., Slota, Jnr., P.J., Gillespie, R., Gowlett, J.A.J., Hedges,

D.R. Harris

R.E.M., Jull, A.J.T., Zabel, T.H., Donahue, D.J. and Berger, R., 1985, Major revisions in the Pleistocene age assignments for North American human skeletons by C-14 accelerator mass spectrometry: none older than 11,000 C-14 years BP, *American Antiquity* 50, 136–140.

Section I

Origins of Agriculture, Plant and Animal Domestication

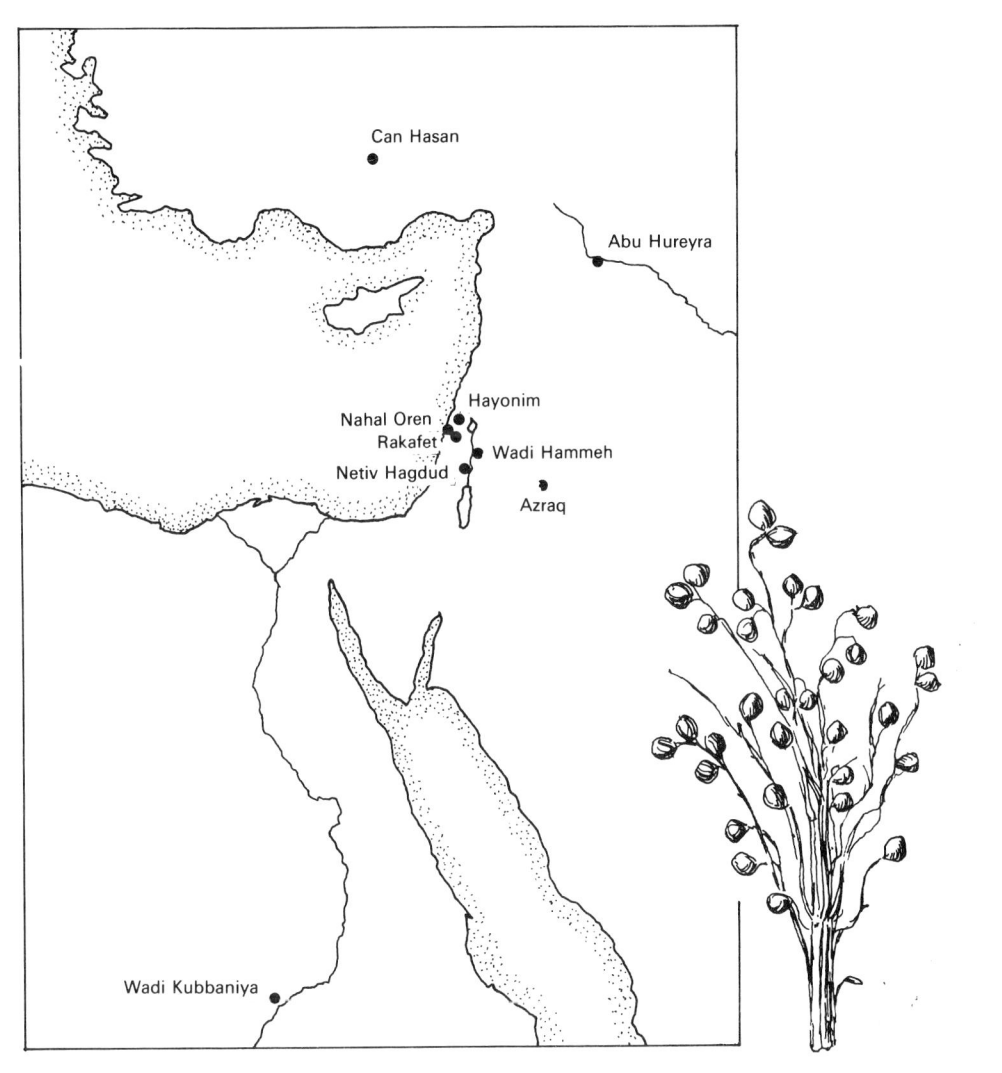

Can Hasan

Abu Hureyra

Hayonim

Nahal Oren

Rakafet

Netiv Hagdud

Wadi Hammeh

Azraq

Wadi Kubbaniya

Overleaf: Sites in the Middle East with accelerator dates; (right) Sorghum (durra) of the 12th century AD from Qasr Ibrim provided by Peter Rowley-Conwy as a known-age sample.

PLANT AND ANIMAL DOMESTICATION AND THE ORIGINS OF AGRICULTURE: THE CONTRIBUTION OF RADIOCARBON ACCELERATOR DATING

D. R. Harris

The question of how, where and when agriculture emerged and became the economic mainstay of human populations remains one of the major unresolved problems of prehistory. Much archaeological endeavour has been directed in recent decades toward elucidating this problem, and many general and regional explanatory models for the emergence of agriculture have been advanced (e.g. Binford 1968; Cohen 1977; Flannery 1968; Harris 1977a, 1977b; Lathrap 1977; Redman 1977; Reed 1977). Great difficulties have been encountered, however, in devising ways of testing such models against archaeological evidence, partly because the models themselves tend to be so highly generalised as to be inherently untestable, and partly because there is too much uncertainty about the character and chronology of the archaeological evidence itself. A major part of that evidence consists of the remains of the plants and animals that were exploited for food and other purposes by prehistoric populations. They include both domesticated and wild forms, the precise identification and dating of which is crucial to a better understanding of the origins and early development of agriculture.

A range of new scientific techniques is now being brought to bear on both the identification and dating of archaeologically recovered plant and animal remains. Such techniques include, for example, the application of scanning electron microscopy, electron spin resonance (Hillman *et el.*, 1983, 1985), nuclear magnetic resonance and pyrolysis mass spectrometry to problems of identification; and, in the realm of dating, the use of radiocarbon accelerator mass spectrometry (AMS). The application of this new dating technique to plant and animal remains is particularly promising because of its capacity to date very small samples. With certain exceptions (such as e.g. large accumulations of animal bones or mollusc shells) plant and animal remains recovered from prehistoric archaeological sites tend to be fragmentary and often few in number. They also have a disconcerting tendency to be stratigraphically mobile after initial deposition, and therefore difficult to relate chronologically, with precision, to their stratigraphic context and to the overall stratigraphy of a site. This problem applies with particular force to small seeds, which are the most ubiquitous class of plant remains directly relevant to studies of agrarian and pre-agrarian plant exploitation. It is highly desirable to be able to date such remains precisely in order to resolve the chronological uncertainty that often undermines confidence in interpretations of plant and animal domestication and the origins of agriculture which are based on unverified stratigraphic associations of different classes of evidence. The advent of AMS dating, with its capacity to date samples as small as a single cereal grain, raises stratigraphic dating to a higher level of precision and provides the archaeologist with a new means of testing many of the chronological assumptions that permeate the literature on agricultural origins. It also means that sampling strategies in this field of research can in future be tailored to the chronological precision that AMS dating offers.

DATING ACHIEVEMENTS

The dating programme so far carried out at the Oxford Radiocarbon Accelerator Unit under the theme of plant and animal domestication and early agriculture consists in large part of what may be described as 'verification dating'. The ability of radiocarbon AMS to cut through chronological confusion by determining whether small, archaeologically controversial samples are as ancient as they are claimed to be has been demonstrated by the Unit in several fields of enquiry, e.g. art history, early man in America, and the origins of agriculture. In particular, the capacity of the technique to test the postulated ages of controversial finds of the remains of domesticated plants and animals has been conclusively shown by a recent series of dates from North African and South-west Asian samples.

The most controversial case is that of the seeds of domesticated barley and date palm recovered at Wadi Kubbaniya in southern Egypt by Fred Wendorf and his colleagues (Wendorf et al. 1979, 1980, 1982, 1984). Excavations of Late Palaeolithic sites in this now-dry tributary of the Nile yielded a variety of plant remains which included, in addition to fragments of wood and charcoal of *Acacia*, *Salsola* and *Tamarix*, seeds identified as of barley, chickpea, lentil and date palm, as well as a single inflorescence fragment of diploid (einkorn) wheat (el Hadidi 1980; Stemler and Falk 1980; Wendorf et al. 1982). The identification of chickpeas and lentils was later shown by Gordon Hillman to be erroneous, but he confirmed that the sample of barley grains included many of the domesticated six-row type. The contexts from which the seeds were recovered were dated by conventional radiocarbon counting of charcoal samples to between 18,500 and 17,000 BP and the excavators were convinced of the stratigraphic integrity of the layers from which the seeds came. However, Hillman's close examination of the barley grains showed that they were not, as had originally been reported, charred, but that the colour of the blackish specimens was probably caused by partial decomposition of the outer cell layers under temporarily wet conditions. As uncharred grains and inflorescence fragments could not have survived at the site if they had been deposited at the time of the Late Palaeolithic occupation, he concluded that the barley grains and einkorn spikelet must be intrusive and that they had been carried down — possibly by ants — through the stratigraphy to the level at which they were found (Hillman 1982, unpublished report to the excavator). This conclusion was later reinforced by the results from electron spin resonance spectroscopy of some of the blackish barley grains which indicated a 'highest past temperature of exposure' insufficient to induce charring (Hillman et al. 1983, 1985).

Because the highly contentious claim that barley (and the other presumed crops) were cultivated at Wadi Kubbaniya in the Late Palaeolithic had been advanced and popularised (Wendorf et al. 1980, pp. 272–279, 1982), it became a matter of high archaeological priority to subject some of the seeds to the independent age test that AMS dating alone could provide. Accordingly, Wendorf submitted to the Laboratory of Isotope Geochemistry at the University of Arizona, Tucson, for AMS dating, six seeds of six-row barley and three small pieces of wood charcoal from the same stratigraphic context as the seeds. He also asked the Oxford Unit to date one of the date-palm seeds and a charcoal sample. The dates obtained at Tucson for the three charcoal samples agreed with those obtained by conventional radiocarbon counting, and thus confirmed the occupation as of Late Palaeolithic age, but none of the barley seeds proved to be contemporary with the occupation. Contamination by tracer ^{14}C in the laboratory, or as a result of their preparation for scanning electron microscope, is thought to have invalidated the results for five of the seeds, and the sixth was found to have a radiocarbon age of 4850 ± 200 years BP (AA-228) (Wendorf *et al.* 1984). The results from Oxford show the date-palm seed also to be intrusive. Separate dates of 350 ± 200 BP (OxA-101) and $101.5\% \pm 2.5\%$ modern standard (OxA-102) were obtained respectively from the insoluble residue and the humic acids of the seed, whereas the charcoal sample gave a date of $17,150 \pm 300$ BP (OxA-103) (Gillespie *et al.* 1984, p. 17).

It is expected that further samples of the plant remains from Wadi Kubbaniya will be submitted for AMS dating, in particular charred specimens of wild plant foods, but meanwhile the ages that have been determined provide an impressive example of the value of verification dating in the investigation of plant domestication and early agriculture. Had the AMS technique not been available to test the contemporaneity of the barley and date seeds with their stratigraphic context, the claim that food production was practised at Wadi Kubbaniya in the Late Palaeolithic would have become established in the literature and might have become widely accepted as archaeological 'fact', with apparently revolutionary implications for our understanding of the origins of agriculture.

The Oxford Unit has also contributed to the resolution of controversial questions about the antiquity of the remains of domesticated crop plants recovered from three pre-Neolithic sites in Israel: Nahal Oren, 'Ain Mallaha and Rakafet. In each case the crop remains, which consisted of wheat grains and, at Rakafet, seed of domesticated bitter vetch, were too scanty to date by the conventional radiocarbon method. Their presence in pre-Neolithic, and therefore (according to conventional interpretations) pre-agrarian contexts, invited verification by AMS dating. The dates obtained for three charred grains of emmer wheat (two domestic and one wild-type) from the Late Palaeolithic (Kebaran) levels at the cave site of Nahal Oren are discussed by Tony Legge, in relation to the complex stratigraphy of the site, elsewhere in this volume. It is sufficient here to stress that the domestic grains proved to be less than 3500 years old.

At the Mesolithic (Natufian) open site of 'Ain Mallaha two grains of domestic wheat were recovered which Hillman identified as from free-threshing wheats of 'bread' and 'macaroni' type which are indicative of long-established agriculture. Dating of this controversial find was therefore given priority. The grain of 'macaroni' type was sacrificed and a date of 330 ± 100 BP (OxA-543) obtained. The seed of domestic bitter vetch recovered from the Late Palaeolithic cave site of Rakafet also proved to be relatively modern, giving a date of 2760 ± 200 BP (OxA-541). Thus the results from the three sites demonstrated that the seeds of domestic type that were found in the pre-Neolithic levels were all intrusive, despite stratigraphic appearances to the contrary, and the putative evidence for Palaeolithic/Mesolithic agriculture was thereby confounded.

As yet the Oxford Unit has undertaken fewer verification dates that relate to animal domestication, although there is considerable potential for dating controversial animal remains. One example of this type of verification dating is the recent determination of the age of a camel jaw bone from the site of Shiqmim in Israel, which was associated with Bronze Age deposits and appeared to pre-date the earliest evidence for domestic camels in South-west Asia. It was dated to 210 ± 150 BP (OxA-135) and must therefore be regarded as intrusive (Gillespie *et al.* 1985, p. 245).

In addition to their achievements in verification dating relating to South-west Asia and North Africa, the Oxford Unit has begun to contribute to the building of a more detailed and precise chronology of changes in prehistoric plant and animal exploitation in that region. The Unit's main contribution so far has been to the dating of plant and animal remains from the Mesolithic (Natufian)-Aceramic Neolithic levels of the site of Tell Abu Hureyra in the Euphrates Valley in northern Syria. This large tell (now drowned under the waters of Lake Assad) was excavated in the early 1970's by Andrew Moore (1975, 1979). An exceptionally large and diverse assemblage of charred plant remains was recovered (Hillman 1975, pp. 70–73) and an equivalently large assemblage of animal bones was also retrieved. Together these bioarchaeological data constitute the most complete sequence presently available of plant and animal exploitation across the chronological threshold from reliance on wild to reliance on domesticated resources. In other words, changes in the plant and animal economy at Abu Hureyra from Mesolithic to Neolithic times are reflected in on-site evidence both of pre-agrarian wild-food procurement and of early agricultural production based on domesticated cereals and livestock (Harris and Hillman 1985; Hillman *et al.* 1986, Legge and Rowley-Conwy, this volume and in press).

The Oxford Unit has so far provided 24 dates on 12 samples of charred cereal grain and charred ungulate bone from the Abu Hureyra sequence: 5 dates on 5 separate samples of wild-type einkorn wheat (3–6 grains per sample), 8 dates on 3 samples of gazelle bone, 7 dates on 3 samples of sheep bone, and 4 dates on 1 sample of cattle bone. The significance of the ungulate-bone dates is discussed elsewhere in this volume by Tony Legge and Peter Rowley-Conwy. The five samples of wild-type einkorn (*Triticum boeoticum*) are from the Mesolithic levels of Trench E at the site and they span the period between *c*. 10,900 and 10,420 BP, as follows: 10,900 ± 200 (OxA-172) Level E326; 10,800 ± 160 (OxA-386) Level E276; 10,600 ± 200 (OxA-170) Level E261; 10,600 ± 200 (OxA-171) Level E313; 10,420 ± 150 (OxA-397) Level E286. The discrepancies that are apparent between the (vertical) sequence of excavated levels (E326 lowest) and the chronological sequence of dates suggest either that there may have been some limited stratigraphic movement of the grain *in situ* or that slight mixing took place during excavation, but all five samples are clearly of Mesolithic age and no intrusion from the higher Aceramic Neolithic levels is indicated. In addition to the wild-type einkorn, a diverse spectrum of wild food plants was exploited during the Mesolithic at Abu Hureyra, whereas it was only in the ensuing Aceramic Neolithic – the evidence suggests – that domesticated cereals and pulses were first cultivated at the site.

The detailed dating of the plant and animal remains from Abu Hureyra is the Oxford Unit's main contribution so far to the investigation of diachronic changes in prehistoric subsistence in South-west Asia. More dates will be run on the Abu Hureyra sequence itself, and other sites in the region which yield evidence of early plant and animal exploitation will be included in the programme. For example, plant remains recently recovered by Susan Colledge from the Mesolithic (Natufian) site of Wadi Hammeh provide some evidence of pre-agrarian exploitation in the Jordan Valley (Edwards *et al.* 1984; Edwards, in press; Edwards and Colledge, in press) – 3 dates on charred seeds of wild leguminous and other herbs from the site have so far been obtained: 11,920 ± 150 BP (OxA-393), 11,950 ± 160 BP (OxA-507) and 12,200 ± 160 BP (OxA-394) – and a series of Late Palaeolithic-Mesolithic (Kebaran-Natufian) sites in the Azraq Basin of east-central Jordan (Wadi el Jilat), which have been sampled by Andrew Garrard and have yielded 11 preliminary charcoal and bone dates, 7 of which span the period between $21,150 \pm 400$ and $11,450 \pm 200$ BP (OxA-519 to Oxa-525 Garrard, pers. comm.), are also expected to yield evidence of pre-agrarian subsistence.

DATING PROSPECTS

It is clear from the foregoing review of some of the initial dating achievements of the Oxford Unit that the capacity of radiocarbon AMS to date very small samples is of great value in investigations of plant and animal domestication and early agriculture. Potentially, it can be applied in many other regional contexts, in addition to South-west Asia, both to verify the postulated ages of specific plant and animal domesticates, and to date closely new pre- and post-agrarian sequences of changing plant and animal exploitation. The Oxford Unit will continue to date South-west Asian samples, but it is now in a position also to contribute to the investigation of early agriculture in other regions of the world that are regarded as possible 'hearth' areas of domestication. An example of its continuing South-west Asian programme is its commitment to date canid bones recovered from the Middle Palaeolithic (Mousterian) site of Douara Cave in central Syria, which have recently been submitted for verification dating by Sebastian Payne. If they prove to be not intrusive but genuinely ancient, then either they represent the earliest domestic dogs known or they are evidence of the existence of a previously unknown population of wild canids which could have played a part in the ancestry of domestic dogs.

Beyond South-west Asia, the Unit has become involved in resolving the chronological uncertainties of the important archaeological sequence of Guitarrero Cave in the Peruvian Andes. This high-altitude site is of particular relevance to studies of plant domestication in tropical America because it has yielded remains of maize, beans, and pepper, as well as tubers tentatively identified as the local cultivars oca (*Oxalis tuberosa*) and ullucu (*Ullucus tuberosus*). The plant remains themselves are not yet available for AMS dating, but the Oxford Unit has dated 15 samples (of charcoal, wooden artefacts, cord and leather) from the site, with results that are in general consistent with the conventional radiocarbon dates previously obtained and which appear to confirm that, although the maize may be less than 2000 or 3000 years old, the beans are probably of equivalent age to the wood and cord samples (*c.* 10,000 years) (Lynch *et al.* 1985, p. 866; Gowlett *et al.* 1986, p. 123).

No other tropical American sites have yet provided samples of plant or animal remains for AMS dating at Oxford, but the technique clearly has the capacity to solve many of the outstanding chronological problems associated with domestication and early agriculture in Middle and South America. For example, systematic application of AMS dating to the plant remains recovered by Richard MacNeish from sites in the Tehuacán Valley in southern Mexico could provide a crucial test of the hypothesis that agriculture developed very gradually there over six or seven millennia (MacNeish 1967, 1972).

In North America, north of the Rio Grande, radiocarbon AMS dating is beginning to be applied to evidence of early plant exploitation and domestication coming from archaeological sites both in the dry South-west and the humid North-east of the United States. This dating is being carried out at the Laboratory of Isotope Geochemistry, Tucson, Arizona. Some preliminary results relating to the North-east have been published (Conard *et al.* 1984), and a review of AMS dates on plant remains from the South-west is being prepared (Long 1985, pers. comm.).

Archaeological sites and sequences that relate to the beginnings of agriculture, the interpretation of which would benefit greatly from the application of AMS dating, exist in many other parts of the world. Foremost among them are sites in South, South-east and East Asia that have yielded remains of rice and other cultivars which remain insecurely dated: e.g. the cave sites in North-west Thailand excavated by Chester Gorman which yielded not only rice but also the remains of some twenty other genera of plants traditionally exploited for food and other domestic purposes and which were dated by the conventional radiocarbon method to the period 11,000–7600 BP (Gorman 1969; Yen 1977). Similarly, finds of rice and other actual or potential cultivars made at other sites in Thailand, as well as elsewhere in South-east Asia, and in China, India and Pakistan, could be incorporated into a programme of AMS dating at Oxford that would make a major contribution to our understanding of early agriculture in tropical Asia.

The same can be said of the potential benefit of applying AMS dating to the investigation of prehistoric subsistence in Europe, Africa, Australasia and the Pacific, where many fundamental chronological questions about the origins and spread of domesticates and agricultural techniques remain unanswered. Nor is the value of the method limited to the direct dating of plant and animal remains. It also has the capacity to provide indirect evidence of past agricultural activity by dating very small sediment samples. This approach to the investigation of early agriculture has yet to be developed, but it holds considerable promise for tackling such refractory problems as the dating of relict field systems. One type of field system now widely recorded in the tropics is that of the raised fields that characteristically occur in lake- and swamp-edge situations and in seasonally flooded lowlands and depressions (Denevan 1970; Parsons 1985; Turner and Denevan 1985). These extensive surface features have proved very difficult to date because they tend to be relatively sterile archaeologically and are seldom directly associated with settlement sites. It may be possible, however, to detect and date evidence of former cultivation by detailed analysis of the stratigraphy, both of the fields themselves and of related sedimentary sequences in adjacent swamp or valley deposits. This approach, which is currently being tried out experimentally on a relict system of agricultural mounds and ditches in the Torres Strait Islands (Barham and Harris 1985; Harris 1985), is likely to depend on AMS dating of very small sediment samples to provide the close chronological control of sedimentary history that is required if the initiation, duration and cessation of cultivation is to be successfully dated.

Overall then, radiocarbon AMS dating has begun to make a significant impact on the investigation of plant and animal domestication and the origins and early development of agriculture. In particular, verification dating has shown how valuable the technique is for exposing errors in the chronological association of stratigraphically mobile plant and animal remains with their depositional contexts, errors which can lead to gross misinterpretations of the evidence and even to the birth of archaeological myths. Verification dating of existing samples is likely to continue for some time to be a most important aspect of AMS dating in this field of study, but, as the stock of uncertain and controversial samples diminishes, it can be expected to become the normal method of dating small bioarchaeological samples and thus greatly to improve chronological precision in the investigation of prehistoric plant and animal exploitation.

REFERENCES

Barham, A.J. and Harris, D.R., 1985, Relict field systems in the Torres Strait region, in *Prehistoric intensive agriculture in the tropics* (ed. I.S. Farrington), pp. 247–283, Oxford: BAR International Series 232.

Binford, L.R., 1968, Post-Pleistocene adaptations, in *New perspectives in archeology* (eds. S.R. Binford and L.R. Binford), pp. 313–341, Chicago: Aldine.

Cohen, M.N., 1977, *The food crisis in prehistory. Overpopulation and the origins of agriculture*, New Haven: Yale University Press.

Conard, N., Asch, D.L., Asch, N.B., Elmore, D., Gove, H., Rubin, M., Brown, J.A., Wiant, M.D., Farnsworth, K.B. and Cook, T.G., 1984, Accelerator radiocarbon dating of evidence for prehistoric horticulture in Illinois, *Nature* 308, 443–446.

Denevan, W.M., 1970, Aboriginal drained-field cultivation in the Americas, *Science* 169, 647–654.

Edwards, P.C., 1984, Two Epipalaeolithic sites in the Wadi Hammeh (Area XX) in Preliminary Report of the University of Sydney's 5th season of excavation (1982–83) at Pella in Jordan (A. McNicoll, W. Ball, S. Bassett, P. Edwards, P. Macumber, D. Petocz, T. Potts, L. Randle, L. Villiers and P. Watson) pp. 55–86, *Annual of the Department of Antiquities of Jordan* XXVIII.

Edwards, P.C., in press, The Epipalaeolithic period, in *Pella in Jordan II* (ed. A. McNicoll), Canberrra: Australian National Gallery.

Edwards, P.C. and Colledge, S.M., in press, The Natufian settlement in the Wadi Hammeh (Area XX), in Preliminary Report of the University of Sydney's 6th season of excavation (1983–84) at Pella in Jordan (T.F. Potts, S.M. Colledge and P.C. Edwards) *Annual of the Department of Antiquities of Jordan.*

el Hadidi, N., 1980, Plant remains from Late Palaeolithic sites in Wadi Kubbaniya, in *Loaves and fishes: the prehistory of Wadi Kubbaniya* (eds. F. Wendorf, R. Schild and A.E. Close), pp. 295–298, Dallas: Southern Methodist University.

Flannery, K.V., 1968, Archeological systems theory and early Mesoamerica, in *Anthropological archeology in the Americas* (ed. B.J. Meggers), pp. 67–86, Washington D.C.: Anthropological Society of Washington.

Gillespie, R., Gowlett, J.A.J., Hall, E.T. and Hedges, R.E.M., 1984, Radiocarbon measurement by accelerator mass spectrometry: an early selection of dates, *Archaeometry* 26, 1, 15–20.

Gillespie, R., Gowlett, J.A.J., Hall, E.T., Hedges R.E.M. and Perry, C., 1985, Radiocarbon dates from the Oxford AMS system: Archaeometry Datelist 2, *Archaeometry* 27, 2, 237–246.

Gorman, C.F., 1969, Hoabinhian: a pebble-tool complex with early plant associations in Southeast Asia, *Science* 163, 671–673.

Gowlett, J.A.J., Hall, E.T., Hedges, R.E.M. and Perry, C., 1986, Radiocarbon dates from the Oxford AMS system: Archaeometry Datelist 3, *Archaeometry* 28, 1, 116–125.

Harris, D.R., 1977a, Alternative pathways toward agriculture, in *Origins of agriculture* (ed. C.A. Reed), pp. 179–243, The Hague: Mouton.

Harris, D.R., 1977b, Settling down: an evolutionary model for the transformation of mobile bands into sedentary communities, in *The evolution of social systems* (eds. J. Friedman and M.J. Rowlands), pp. 401–417, London: Duckworth.

Harris, D.R., 1985, Palaeoenvironmental methods for dating prehistoric field systems in seasonally flooded tropical wetlands, Research project funded by the Science-based Archaeology Committee of the Science and Engineering Research Council. Typescript.

Harris, D.R. and Hillman, G.C., 1985, Early manipulations of plant resources in Near Eastern steppe and riverine forest, Final Report to the Science-based Archaeology Committee of the Science and Engineering Research Council. Typescript.

Hillman, G.C., 1975, The plant remains from Tell Abu Hureyra: a preliminary report, in The excavation of Tell Abu Hureyra in Syria: a preliminary report, *Proc. Prehist. Soc.* 41, 50–77.

Hillman, G.C., Colledge, S.M. and Harris, D.R., 1986, Plant-food economy during the Epi-Palaeolithic period at Tell Abu Hureyra, Syria: dietary diversity, seasonality and modes of exploitation. Pre-circulated paper for the Symposium on Recent Advances in the Understanding of Plant Domestication and Early Agriculture, World Archaeological Congress, Southampton, September 1986.

Hillman, G.C., Robins, G.V., Oduwole, D., Sales, K.D. and McNeil, D.A.C., 1983, Determination of thermal histories of archaeological cereal grains with electron spin resonance spectroscopy, *Science* 222, 1235–1236.

Hillman, G.C., Robins, G.V., Oduwole, D., Sales, K.D. and McNeil, D.A.C., 1985, The use of electron spin resonance spectroscopy to determine the thermal histories of cereal grains, *J. Archaeol. Sci.* 12, 49–58.

Lathrap, D.W., 1977, Our father the cayman, our mother the gourd: Spinden revisited, or a unitary model for the emergence of agriculture in the New World, in *Origins of agriculture* (ed. C.A. Reed), pp. 713–751, The Hague: Mouton.

Legge, A.J. 1986, Seeds of discontent: accelerator dates on some charred plant remains from the Kebaran and Natufian cultures, this volume.

Legge, A.J. and Rowley-Conwy, P.A., in press, Steppe hunters and the rise of domestication: the evidence from Tell Abu Hureyra, *Scientific American*.

Lynch, T.F., Gillespie, R., Gowlett, J.A.J. and Hedges, R.E.M., 1985, Chronology of Guitarerro Cave, Peru, *Science* 229, 864–867.

MacNeish, R.S., 1967, A summary of subsistence, in *The prehistory of the Tehuacan Valley, Vol. 1: Environment and subsistence* (ed. D.S. Byers), pp. 290–309, Austin: University of Texas Press.

MacNeish, R.S., 1972, The evolution of community patterns in the Tehuacan Valley of Mexico and speculations about the cultural processes, in *Man, settlement and urbanism* (eds. P.J. Ucko, R. Tringham and G.W. Dimbleby), pp. 67–93, London: Duckworth.

Moore, A.M.T., 1975, The excavation of Tell Abu Hureyra in Syria: a preliminary report, *Proc. Prehist. Soc.* 41, 50–77.

Moore, A.M.T., 1979, A pre-Neolithic farmers' village on the Euphrates, *Scientific American* 241, 50–58.

Parsons, J.J., 1985, Raised field farmers as pre-Columbian landscape engineers: looking north from the San Jorge (Colombia), in *Prehistoric intensive agriculture in the tropics* (ed. I.S. Farrington), pp. 149–165, BAR International Series 232.

Redman, C.L., 1977, Man, domestication, and culture in southwestern Asia, in *Origins of agriculture* (ed. C.A. Reed), pp. 523–541, The Hague: Mouton.

Reed, C.A., 1977, A model for the origin of agriculture in the Near East, in *Origins of agriculture* (ed. C.A. Reed), pp. 543–567, The Hague: Mouton.

Stemler, A.B.L. and Falk, R.H., 1980, A scanning electron microscopic study of cereal grains from Wadi Kubbaniya, in *Loaves and fishes: the prehistory of Wadi Kubbaniya* (eds. F. Wendorf, R. Schild, and A.E. Close), pp. 299–306, Dallas: Southern Methodist University.

Turner, B.L. and Denevan, W.M., 1985, Prehistoric manipulation of wetlands in the Americas: a raised field perspective, in *Prehistoric intensive agriculture in the tropics* (ed. I.S. Farrington), pp. 11–30, BAR International Series 232.

Wendorf, F., Schild, R., el Hadidi, N., Close, A.E., Kobusiewicz, M., Wieckowska, H., Issawi, B. and Haas, H., 1979, Use of barley in the Egyptian Late Palaeolithic, *Science* 205, 1341–1347.

Wendorf, F., Schild, R. and Close, A.E. (eds.), 1980, *Loaves and fishes: the prehistory of Wadi Kubbaniya*, Dallas: Southern Methodist University.

Wendorf, F., Schild, R. and Close, A.E., 1982, An ancient harvest on the Nile, *Science 82*, 3, 68–73.

Wendorf, F., Schild, R., Close, A.E., Donahue, D.J., Jull, A.J.T., Zabel, T.H., Wieckowska, H., Kobusiewicz, M., Issawi, B. and el Hadidi, N., 1984, New radiocarbon dates on the cereals from Wadi Kubbaniya, *Science* 225, 645–646.

Yen, D.E., 1977, Hoabinhian horticulture? The evidence and the questions from Northwest Thailand, in *Sunda and Sahul. Prehistoric studies in Southeast Asia, Melanesia and Australia* (eds. J. Allen, J. Golson and R. Jones), pp. 567–599, London: Academic Press.

SEEDS OF DISCONTENT: ACCELERATOR DATES ON SOME CHARRED PLANT REMAINS FROM THE KEBARAN AND NATUFIAN CULTURES

A. J. Legge

INTRODUCTION

From the time of its first description (Garrod and Bate 1937, Garrod 1957), the Natufian Culture of Palestine has been regarded as a satisfactory cultural precursor for the emergence of settled communities based upon full-scale agriculture and animal husbandry. The reasons for this were largely based upon the artefacts characteristic of the Natufian; elaborate graves, art work, microlithic tools, edge-lustered sickles and the large conically-pierced 'mortars' all provided the setting of technological intensification that was anticipated for a proto-agricultural community. Besides such expectations, the sites were known to be within or near the area of distribution of wild emmer wheat, *Triticum dicoccoides*. Yet in spite of this, almost no attempt had been made to recover the material (seeds preserved by charring or other means) that was needed to investigate the economic importance of plants. Before various water flotation systems were developed for the recovery of charred plant remains, only highly visible samples were saved, such as the charred fills of pits or hearths, and such deposits are not characteristic of Natufian sites.

The work of Helbaek on plant remains recovered from Jarmo (Braidwood and Howe 1960) in Iraq provided the first detailed overview of the plant remains from an early farming village in the Near East. The study was based upon material preserved both in the charred state and as impressions in *tauf*. Helbaek again was responsible for the study of plant remains in the Deh Luran region of Iran, from excavations by Hole, Flannery and Neely (1969). In this report, more attention was given to the description of the retrieval methods used for both bone and plant remains; for this latter class of material, water flotation was employed, with the method being derived from work done in America by Struever (1968).

While such methods can give satisfactory results, investigations by the author and others suggested that a technique of flotation was needed that operated on a larger scale and allowed the continuous processing of a large volume of sediment combined with the use of wet-sieving with fine meshes of 0.2 or 0.3 mm (Payne 1971). A number of research staff in the British Academy early agriculture research project, then based in the Department of Archaeology at Cambridge University, applied considerable effort to the development of such a technique, and to its field application at archaeological sites in Britain, in mainland Europe and the Near East. Among the sites investigated were the cave of Nahal Oren (Wadi Fellah) (Noy *et al.* 1973) and the open air settlement of 'Ain Mallaha (Perrot 1960).

13

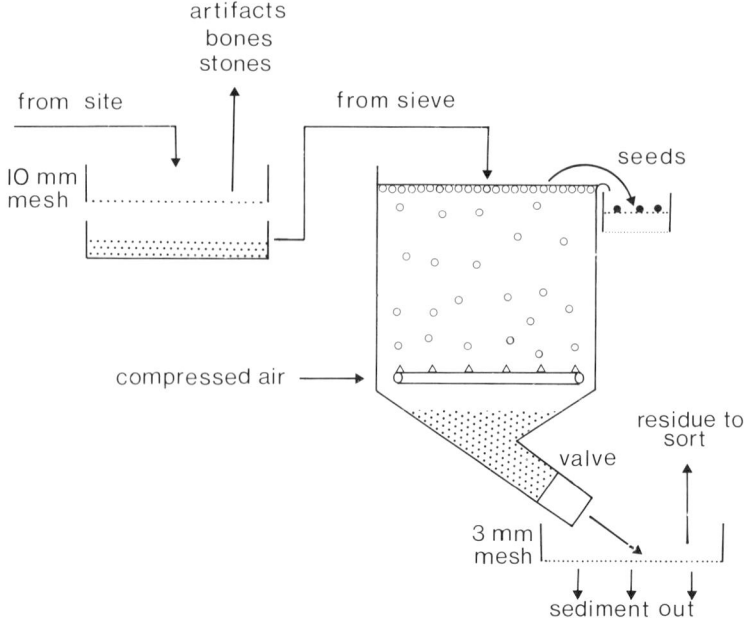

Fig. 1   Flotation system used at Nahal Oren

METHOD

The 'Cambridge' flotation system that was developed for the recovery of plant remains was based upon the industrial technique of froth flotation (Jarman *et al.* 1972). A tank containing water had compressed air fed into its base, distributed as fine bubbles through porous bronze filter cones. The water contained a small amount of a 'collector' (a hydrocarbon — usually paraffin) that was chosen for its affinity with the material to be recovered. This was intended to render the desired material (charred seeds and other charred plant structures) water repellent, so that the bubbles would adhere to the seed surface and aid its lifting into the froth bed. As the sediment displaced water in the flotation tank the froth bed overflowed a weir, and the seeds were carried to fine-mesh collecting sieves.

In running the process, soil was dry sieved through a 1 cm mesh, placed in plastic buckets, and then was poured in a steady stream into the froth bed on the surface of the water tank. The rising bubbles acted in breaking down the crumb structure of the sediment and assisted in raising the seeds to the surface. The mineral sediment sank into the cone-shaped base of the tank from where it was released by a valve as mud, into the 'wet' sieves. All sediment volumes were recorded, firstly by the number of buckets delivered to the 1 cm 'dry' sieve, and then by the quantity of sieved residue (minus the larger stones, implements and bones) that was passed to the flotation machine. Processing was therefore continuous as long as the labels on the buckets remained unchanged in terms of their layer or feature numbers. To avoid the possibility of cross-contamination when a new batch of sediment was started, the tank was drained, cleaned internally, and the water returned after filtration through a mesh of 300 microns. The process is shown in diagrammatic form in Fig. 1.

PLANT REMAINS AND STRATIGRAPHY OF NAHAL OREN

All sediments were excavated by trowel and were then dry sieved, processed by flotation and wet sieved as described above. Plant remains were recovered from all of the trenches excavated. In all 133 seeds or fruits were found that were unquestionably preserved by charring. A preliminary list of identifications was published by Legge and Dennell (1973). Most attention has been given to the grains of *Triticum sp.* from the earlier levels of the site, though it should be remembered that these are associated with a larger number of seeds from the tribe *Viciae*, which are yet to be dated.

The grains of *Triticum* with which this paper is concerned come from the trench designated as 300. This was located on the uppermost part of the talus, adjacent to a wall built across the cave mouth to enclose goats (Noy, Legge and Higgs 1973, Fig. 1, p. 76). The trench was excavated to bedrock, through about 3.0 m of well stratified sediments, which follow the dip of the bedrock at 20°–25° from the horizontal. At the downslope end of trench 300 a massive rockfall separated the stratigraphy from the adjacent trench 500 immediately below. This rockfall occurred during the Kebaran occupation of the site (Fig. 2) when some sediment had accumulated; the rockfall lies about 20–30 cm above the bedrock.

Trench 300 has a stratification much as the other deep trenches on the talus with Pre-Pottery Neolithic beginning at the modern ground surface. Layers of Natufian and then

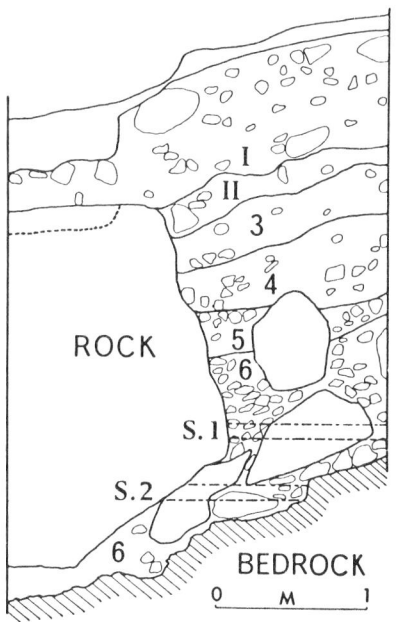

Fig. 2 Section of trench 300 at Nahal Oren

Kebaran occupation continue in an unbroken sequence down to the bedrock. The trench was excavated in 9 main layers which were later amalgamated into 6, as divisions had been made during excavation by the degree of brecciation within layers otherwise undifferentiated by colour or stone content. In the recording system employed for the excavation of Nahal Oren, a label such as 308.0/28 indicates that layer 8 of trench 300 was being excavated, with no particular feature being differentiated within the layer (i.e. the zero is the 'feature number', not used in this example). The figure after the stroke indicates depth: in this case, 2.80 m below the highest point of the trench sections. Although the original record numbers are retained here, all of the *Triticum* grains from trench 300 come from the Kebaran layer 6 as shown in Fig. 2.

From the record made during excavation, this layer was described as extremely hard, stony and brecciated. The site notebooks also contain records of sediment volumes processed and, for the level containing the *Triticum* grains, this amounts to 0.680 cu.m. taken to be dry sieved. When all particles of > 1 cm in size were removed by dry sieving, only 70% of this volume remained to be processed by flotation. For sample 307.0/28, about 24% of the sediment volume was removed as stones, flints and bones; for 307.0/33 this was 33% and for sample 308.0/32–33, it was 39%. The stones in these levels were mainly fist-sized limestone fragments, with irregular weathered surfaces. Very few rounded pebbles from the wadi below were found.

Charcoal, apart from the charred seeds found, was sparse at Nahal Oren, and radiocarbon dates were therefore done on charred bone, selected from the same trench and levels as the charred *Triticum*. The mammal bone from almost all layers was very comminuted. This was due in part to the small body size of the prevailing mammal that was hunted (*Gazella*) but also to extreme bone processing by the inhabitants of the site. Bone from the layer was also heavily indurated, and required treatment with dilute acetic acid before identifications could be made. Accordingly, the larger splinters from the diaphyses of limb bones were tested *before* acid treatment by breaking off a small part of one end; those showing evidence of charring were set aside as potential radiocarbon samples. The radiocarbon dates were done by Dr R. Protsch at the Institute of Geophysics, University of California, as part of a programme of dating early Near Eastern mammal bone (Protsch and Berger 1973). These dates can be considered in relation to the new accelerator dates performed on the grains of *Triticum* from the same levels.

THE CHARRED TRITICUM

OxA-390 Specimen 307.0/32–33   > 33,000 years.

This specimen comes from immediately below the level dated at $16,880 \pm 340$ BP (UCLA-1776B; Fig. 2, sample S2); the label indicates that a small amount of level 307.0/32 was incorporated into the sample.

OxA-389 Specimen 307.0/33 2940 ± 120 BP

This is from the same level as the charred bone sample, dated at $16,880 \pm 340$ BP (UCLA-1776B, Fig. 2, S2).

OxA-395 Specimen 308.0/28 (charred material) 3100 ± 130 BP
OxA-396 Specimen 308.0/28 (humic extract) 6650 ± 190 BP

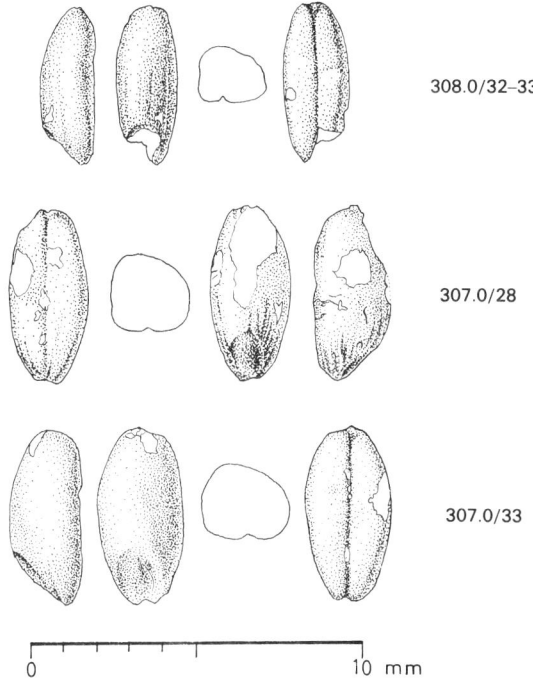

308.0/32–33

307.0/28

307.0/33

0   10 mm

Fig. 3   Charred Triticum grains from Nahal Oren

This lies immediately above charred bone sample 307.0/29, dated by conventional radiocarbon at 15,800 ± 300 BP (UCLA-1776A; Fig. 2, S1).

A third date made upon charred bone came from trench 400 (Noy *et al.* 1973, Fig. 1) which was situated at the lowest point of the surviving talus, in order to utilise the recent road cutting as a section, so that the deepest accessible part of the stratigraphy could be sampled. The Kebaran layer IX in this trench gave a date of 18,250 ± 320 BP (UCLA-1776C).

The UCLA dates fall in the correct stratigraphic order, and appear to be of an appropriate magnitude for the culture concerned. Although there are relatively few dates available for the more northerly Kebaran sites, those available do not greatly differ from Nahal Oren (Weinstein 1984). However, the point here is not a discussion of Kebaran chronology, but rather to indicate that the dates are acceptable. To set against this, the seed dates from the Oxford accelerator are *either* substantially older *or* younger than the charred bone dates. The *Triticum* specimens dated by accelerator are shown in Fig. 3, as drawn by Richard Hubbard of the North-East London Polytechnic. One specimen (307.0/32–33) shows the narrow, almost parallel-sided form, with low dorsal surface and strong compression lines imprinted by the lemma, typical of grains of the Syro-Palestinian races of wild emmer *Triticum dicoccoides* sub sp. Syrico-palestinicum (G. Hillman, pers. comm.). This specimen is also the most ancient (> 33,000 BP). The other two seeds are much fatter and more rounded, and show all of the hallmarks of typical domestic emmer *Triticum dicoccum*, as was suggested by Legge and Dennell (1973). In the years since the discovery of these specimens the desirability of more precise dates had been increasingly felt. With the advent of the accelerator method of radiocarbon dating, this has become possible due to the small mass of ancient carbon required for a sample (Hedges and Gowlett 1986). The dates show the charred seeds to be either very ancient or much more recently intrusive. How can this be interpreted?

Firstly, specimen 307.0/32–33 is about twice as old as its cultural and stratigraphic context would imply. Given that this date is correct (and there seems little reason for it to be rejected) this is certainly the oldest specimen of *Triticum dicoccoides* known, though it must be regarded as derived from more ancient sediments at the site. Nahal Oren has earlier Palaeolithic occupation, though none survives upon the bedrock outside the cave mouth. The excavation of trench 400 near to the foot of the talus showed the presence of earlier Upper Palaeolithic industries below the Kebaran layers, although the dangers from road traffic prevented the full exploration of this trench. With the presence of these earlier cultures at Nahal Oren, the most likely provenance for this specimen must be by derivation from the erosion or re-deposition of material from older archaeological levels. For the others, intrusion is the only possible explanation.

Contamination of the samples during flotation can be discounted as a possible source. No sediments were handled on the site that may have been the source of the grains dated to about 1000 bc. Visible intrusions into the sediments at Nahal Oren were also rare. A few sherds of pottery were found within the archaeological sediments of trench 300, but none of these had penetrated more than 20–30 cm from the modern surface. However, a house near to the site was occupied by a family of herders at the time of excavation, and traces of an earlier (but not prehistoric) house were found upon which the goat retaining wall had been built. It would seem that the site was occupied sporadically during the historic period, and charred grain would logically form part of the wastes discarded from such a house.

Where any fine-grained sediments are exposed to prolonged drying, cracking may occur, possibly initiated at a vertical rock face. In the excavation of the Hotu cave in Iran, Coon (1952) described a disturbed zone with air pockets against the rockface in the upper three metres of the cave fill, though in relatively unconsolidated sediments of historic date. Such a zone of disturbance was not seen in the sections at Nahal Oren and, with a slope of more than 20° in the bedding of the sediments above the rock, it was felt that such a fissuring was unlikely due to pressure from the sediments resting against the vertical face of the rockfall. In three summmers of excavation during July to late September such fissures were not seen to develop.

An alternative possibility for this contamination lies in the activities of burrowing animals. Large burrows, such as those of rodents, are occasionally encountered in excavating ancient sediments. Only those which are still open are likely to be visible as, unless infilled with sediment of a very different colour, when collapsed all trace is likely to be lost. Less visible are the activities of ants. In the Near East, these are known to burrow extensively, and also to engage in the large-scale underground storage of seeds. This was described by the traveller Charles Doughty.

> "We rode by a threshing ground; . . . I saw the emmet's [i.e. the ant's] last confusion (which they suffered as robbers' — their hill-colonies subverted, and caught up in the women's meal sieves! . . . And each needy wife had already some handfuls laid up in her spread kerchief." (abbreviated from Doughty 1936, II, pp. 417–418)

However, Doughty was recording the theft of fresh grain, not that which was charred. While it seems probable that ants were responsible for the introduction of uncharred grain, now known to be of recent origin (Hillman *et al.* 1983, Wendorf *et al.* 1984), into late Palaeolithic sites in Egypt (Wendorf *et al.* 1979, 1984), it seems improbable that they would actively store charred grains. While partly charred seeds might be selected by ants, those from Nahal Oren were heavily charred, to the point where they showed the characteristic vesicular internal structure which results from this treatment. Whatever the mechanism, the vertical intrusion of charred grain into deep and seemingly well-sealed deposits is a problem that is now more widely recognised. For example, Hillman (1982) found a typical Medieval assemblage of cereals and pulses intruding through deep sediments, into Mesolithic deposits at a site in North Wales.

'AIN MALLAHA (PERROT 1960)

At the invitation of Dr Jean Perrot, sediment samples were processed by flotation from level 3 within a round house of the Natufian period at this site which was under excavation during 1971. A small amount of charred material was recovered, including two poorly preserved but very rounded grains of *Triticum*. The presence of these specimens was noted in a paper by Legge (1977), but they have not otherwise been published. The specimens were examined by Gordon Hillman in 1982, who identified the more intact specimen as a free threshing wheat, probably a short-grained form of macaroni or Rivet wheat, *Triticum durum* or *Triticum turgidum*. The second, less well preserved grain was identified as "some form of bread wheat, *Triticum aestivum* s.1." Such wheats are not only fully domesticated, but also represent advanced races which otherwise do not make their appearance before the 7th millenium bc, for example at the site of Can Hasan III (Hillman 1973). The better preserved grain has now been submitted for an accelerator date, and has also proved to be of recent origin:

OxA-543 *Triticum aestivum*   330 ± 100 BP

CONCLUSIONS

Accelerator dates on charred *Triticum* grains at Nahal Oren have shown that such material can intrude more than three metres below the ground surface, and into a stony, consolidated talus deposit. Such intrusions of very small objects must arise from disturbance caused by a combination of rodent and insect burrows, and natural ground movements from the contraction and expansion of sediments in alternate wetting and drying. Yet in the same archaeological levels, specimens had apparently been derived from much more ancient sediments, and had survived in good condition. This shows that inspection alone cannot reveal the differences between very ancient and much more recent charred seeds. The Nahal Oren *Triticum* appeared acceptable at the time of its discovery on the grounds of a lack of visible contamination in the sediments, on the associated fauna, the stratigraphy and the consistent radiocarbon dates done on charred bone. However, it is evident that material from such talus deposits must be regarded with more caution and, for all ancient seeds (especially from those sites with sparse seed remains), accelerator dates will play an essential role in proving them to be contemporary with their sediments.

The impact of these findings on our understanding of the ancient food economy at Nahal Oren is limited. Even had all the *Triticum* grains proved to be ancient, there is still little evidence that cereals were an abundant food for people of the Kebaran and Natufian cultures. The one genuinely ancient grain shows that such material can survive in stony cave sediments and, bearing in mind the great abundance of charred seeds in contemporary sediments at Tell Abu Hureyra (Hillman 1975), the rarity of *Cerealia* at Nahal Oren may not be a factor of preservation alone, but a genuine expression of infrequent use. Further samples (especially of the tribe *Viciae*) from Nahal Oren await accelerator dating, and these may throw a little more light on the question of pre-Neolithic plant use at these sites.

ACKNOWLEDGEMENTS

I am indebted to Richard Hubbard of the North-East London Polytechnic for his drawings of the specimens. Gordon Hillman described the specimens before dating, and I am grateful for his permission to quote from his notes here.

REFERENCES

Braidwood, R.J. and Howe, B., 1960, *Prehistoric Investigations in Iraqi Kurdistan*, Studies in Oriental Civilisation 3l, Chicago: University of Chicago Press.

Coon, C.S., 1952, Excavations in Hotu Cave, Iran: a preliminary report, *Proc. of the American Philosophical Soc.* 96, 3, 231–257.

Doughty, C.M., 1936, *Travels in Arabia Deserta*, New York: Dover Publications.

Garrod, D.A.E., 1957, The Natufian Culture: life and economy of a Mesolithic people in the Near East, *Proc. of the British Academy* 43, 211 ff.

Garrod, D.A.E. and Bate, D.M.A., 1937, *The Stone Age of Mount Carmel, I*, Oxford: O.U.P.

Hedges, R.E.M. and Gowlett, J.A.J., 1986, Radiocarbon dating by accelerator mass spectrometry, *Scientific American* 254, 1, 100–107.

Hillman, G.C., 1973, The plant remains, in Excavations at Can Hasan III 1969–1970 (D.H. French, G.C. Hillman, S. Payne and R.J. Payne) in *Papers in Economic Prehistory* (ed. E.S. Higgs), pp. 183–190, Cambridge: The University Press.

Hillman, G.C., 1975, The plant remains, in The Excavation of Tell Abu Hureyra in Syria: a preliminary report, *Proc. Prehist. Soc.* 41, 50–69.

Hillman, G.C., 1982, Charred remains of Medieval crops from Mesolithic levels: an example of vertical intrusion into deep deposits, Appendix to Excavations at Hendre (Rhuddlan): the Mesolithic finds (J. Manley and E. Healey), *Archaeologica Cambrensis* 131, 43–44.

Hillman, G.C., Robins, G.V., Oduwole, D., Sales, K.D. and McNeil, D.A.C., 1983, Determination of thermal histories of archaeological cereal grains with electron spin resonance spectroscopy, *Science* 222, 1235–1236.

Hole, F., Flannery, K.V. and Neely, J., 1969, *Human ecology in the Deh Luran Plain*, Memoirs of the Museum of Anthropology, Ann Arbor, Michigan, No. 1.

Jarman, H.N., Legge, A.J. and Charles, J.C., 1972, The retrieval of plant remains from archaeological sites by means of froth flotation, in *Papers in economic prehistory* (ed. E.S. Higgs), pp. 39–49, Cambridge: The University Press.

Legge, A.J., 1977, The origins of agriculture in the Near East, in *Hunters, gatherers and first farmers beyond Europe* (ed. V. Megaw), Leicester: The University Press.

Legge, A.J. and Dennell, R.W., 1973, Plant remains, in Recent excavations at Nahal Oren, Israel, (T. Noy, A.J. Legge and E.S. Higgs), *Proc. Prehist. Soc.* 39, 75–99.

Noy, T., Legge, A.J. and Higgs, E.S., 1973, Recent excavations at Nahal Oren, Israel, *Proc. Prehist. Soc.* 39, 75–99.

Payne, S., 1971, Partial recovery and sample bias: the results of some sieving experiments, in *Papers in economic prehistory* (ed. E.S. Higgs), pp. 49–64, Cambridge: The University Press.

Perrot, J., 1960, Le gisement Natoufien de Mallaha (Eynan), Israel, l'*Anthropologie* 70, 437–483.

Protsch, R. and Berger, R., 1973, Earliest radiocarbon dates for domesticated animals, *Science* 179, 235–239.

Struever, S., 1968, Flotation techniques for the recovery of small-scale archaeological remains, *American Antiquity* 33, 353–362.

Weinstein, J.M., 1984, Radiocarbon dating in the southern Levant, *Radiocarbon* 26, 3, 297–366.

Wendorf, F., Schild, R., El Hadidi, N., Close, A., Kobusiewicz, M., Wieckowska, H., Issawi, B. and Haas, H., 1979, Use of barley in the Egyptian late Palaeolithic, *Science* 205, 1341–1347.

Wendorf, F., Schild, R., Close, A., Donahue, D.J., Jull, A.J.T., Zabel, T.H., Wieckowska, H., Kobusiewicz, M., Isaawi, B. and Hadidi, N., 1984, New radiocarbon dates on the cereals from Wadi Kubbaniya, *Science* 225, 645–646.

NEW RADIOCARBON DATES FOR EARLY SHEEP
AT TELL ABU HUREYRA, SYRIA

A. J. Legge and P. Rowley-Conwy

INTRODUCTION

Although sheep are now the world's most abundant domesticated mammal, their remains are rare in the archaeological sites of the Near East before the middle of the eighth millennium bc. Little is known of the distribution of wild sheep in the late Pleistocene (Payne 1968), nor is it possible at this stage to assign the remains that we have to a particular sub-species. New finds of sheep bones from Epipalaeolithic layers at Tell Abu Hureyra in Syria have been directly dated by the accelerator method at the University of Oxford (Hedges and Gowlett 1986), showing that they are indeed contemporary with the earliest settlement at that site. This new material, combined with recent finds reported by Davis *et al.* (1982), Davis (1985) and Payne (1983), extend the range of early wild sheep to a steppe environment considerably to the south and west of their known distribution in the Elburz and Zagros Mountains.

In the mountain regions of Anatolia and Iran the modern species of wild sheep are the Mouflon, *Ovis ammon orientalis* (chromosome number 2n = 54) in the west of the region (Anatolia and north-western Iran) and the Urial, *Ovis ammon vignei* (2n = 58) in north-eastern Iran and east to Kazakhstan, Afghanistan and the foothills of the Himalayas (Nadler *et al.* 1971, 1972, 1973). In the eastern Elburz Mountains, at the southern end of the Caspian Sea, the two species intergrade. The 2n = 54 chromosomes of the Mouflon group grade through animals with 2n = 55, 56 and 57 chromosomes to the 2n = 58 of the Urial group. Horn and coat characteristics in the males change broadly in association with chromosome number, from the simple supracervical curve in horns of the Mouflon to a full circle curve with everted tips of the Urial. Modern domestic sheep have a chromosome number of 2n = 54 and, according to Nadler *et al.* (1971, 1973), the karyotype of domestic sheep is most closely related to that of the western Mouflon group of wild sheep.

Yet the depression of the snowline during the severe late Pleistocene climate must have caused a different distribution of wild sheep from that seen at the present day. In consequence, it would be unwise to attribute a sub-species name to much of the archaeological remains that we have, but rather to assign them to the general species of *Ovis ammon*.

SHEEP BONES FROM TELL ABU HUREYRA

At Tell Abu Hureyra in Syria, the stratigraphy is exceptional, with extensive Epipalaeolithic levels beneath those of the early Aceramic Neolithic (Moore *et al.* 1975), a feature shared by very few large tells. Sheep and goats were found to be present from the earliest levels at the site (Legge 1975) and, with the more extensive work on the mammal remains completed in recent years, it is evident that both species are represented in the Epipalaeolithic in about equal numbers (Legge and Rowley-Conwy, forthcoming). However, they form only a small part of the fauna, at about 6–8%. The mammalian remains associated with the sheep bones are overwhelmingly of gazelles, mainly of the species *Gazella subgutturosa*. Water dependent species such as deer, pig and — at least for the very dry part of the year — cattle, are rare. In the early Aceramic Neolithic settlement at Abu Hureyra the fauna continues very much the same as that of the Epipalaeolithic, Steppe species predominate, and the only significant change in the fauna, associated at this point with the appearance of established plant husbandry (G. Hillman, pers. comm.), is the virtual disappearance of the hare (*Lepus capensis*).

During the earlier Aceramic Neolithic, sheep and goats show some increase over their low frequency in the Epipalaeolithic. Dividing the archaeological levels into a series of sequential units, each containing *c.* 1000 bones, shows that the proportion of sheep and goat fluctuates from 8–16%. Although our numerical analyses are incomplete, an initial examination of the data shows that sheep increase at the expense of goats through the Abu Hureyra sequence. In the later Aceramic Neolithic, the proportions of sheep-goat and gazelles show, in terms of an archaeological time scale, a fairly abrupt reversal. The gazelles, formerly abundant, fall to about 20% of the mammalian remains, while sheep and goat increase to more than 80%. Among the larger mammals at Abu Hureyra, this represents a major shift in the economy on a scale not before seen. While there were evident changes before this time the scale was, in comparison, minor.

ACCELERATOR DATES UPON SHEEP BONES AT TELL ABU HUREYRA

The sequence at Tell Abu Hureyra can be broadly viewed as a gazelle-based economy persisting through the Epipalaeolithic and Aceramic Neolithic in which sheep and goat were a minor component, while later levels have abundant sheep and goat. Obviously, the stratigraphic disturbances associated with settlement during the accumulation of tell sediments, such as the digging of pits, and graves, or the levelling of surfaces for house building and so on, render the cross-examination of earlier archaeological levels a probability to some degree. Thus our concern was to know whether the bones of sheep and goat in the early levels at the site may have been introduced from later levels above by disturbance. Such problems are commonplace in archaeology, and affect a wide range of different artefacts. Indeed, many of the sheep bones from the Palaeolithic sites described below come from caves where occupation has been sporadic up to recent times; from certain of these sites the excavators have reported potsherds and metal objects intrusive even into Middle Palaeolithic levels.

To set against the possibility of intrusion, the large size of the sheep bones reported from sites such as Shanidar and Bisitun (Uerpmann 1978, 1979) give considerable weight to the early dates implied by the stratigraphic context; were modern or historic sheep intrusive, they would probably be of much smaller body size. However, at Tell Abu Hureyra, any sheep bones that were intrusive would not be from large late Pleistocene sheep, but from those of the Aceramic Neolithic, which would be little different in size and at most one or two millennia younger in date. Accordingly, suitable bones were chosen so that this problem might be tested by radiocarbon dating.

In conventional radiocarbon dating, mammal bones from the Near East may present

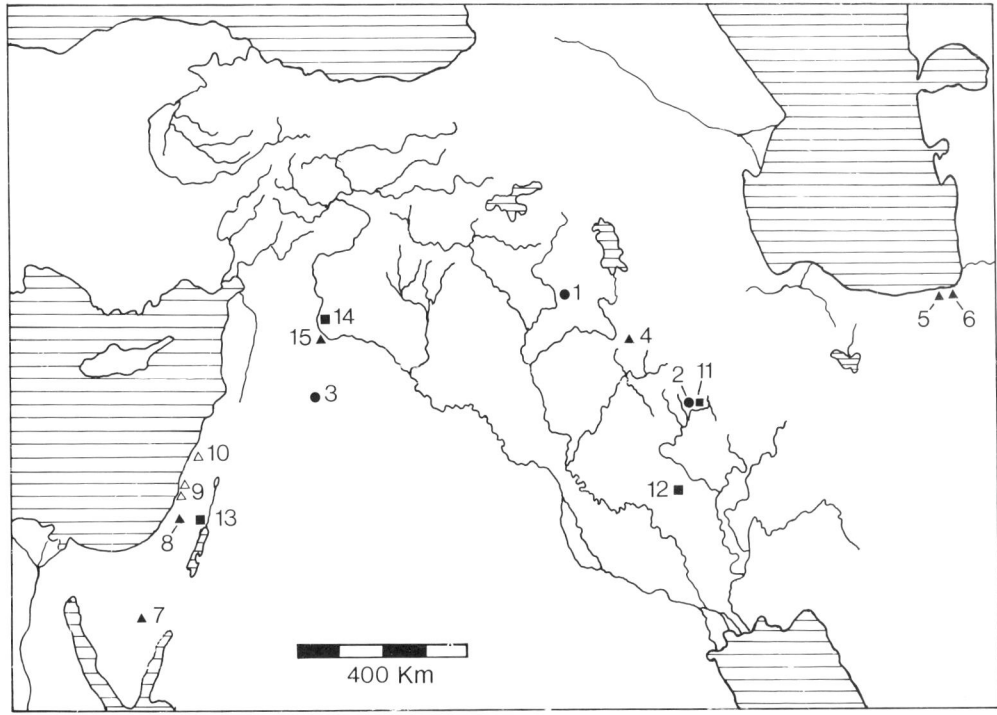

Fig. 1   Near Eastern sites with early sheep remains

● *Middle Palaeolithic finds*
1   Shanidar cave
2   Bisitun cave and Warwasi rock shelter
3   Douara cave
▲ *Upper/Epipalaeolithic finds*
4   Palegawra cave
5   Belt and Hotu caves
6   Ali Tappeh cave
7   Rosh Horesha, Ramat Harif, Abu Salem
8   Hatoula
15 Tell Abu Hureyra

△ *Sites with no sheep bones reported*
9   Kebarah cave
10 Wadi Mughara caves, Nahal Oren (Wadi Fellah)
■ *Aceramic Neolithic finds*
11 Tepe Asiab, Tepe Sarab, Ganj Fellah
12 Tepe Ali Kosh
13 Tell es-Sultan (Jericho)
14 Tell Mureybit

particular problems. In a project concerned with dating early domestication, Protsch and Berger (1973) utilised collagen extracted from bones of cattle, pig, sheep and goat, mainly from sites in Iran. Yet in material from Tell Abu Hureyra the bone collagen has been wholly lost. In such a case, burned bone is the obvious choice in that the charred collagen, being resistant to chemical or bacterial destruction, remains in the bone to be dated. For a date done by the conventional method, about 100g of such charred material would be needed. Bones from a site such as Abu Hureyra are also, in general, very comminuted. This presents a number of problems.

Firstly, all but the largest pieces of bone would be insufficient for such testing unless the sample was bulked by combining many fragments; a substantial number of bones, all of them certainly identified as sheep, would then be destroyed to provide the date. The loss would be serious as a substantial part of the sample from the earliest levels at the site would have to be destroyed. Secondly, the very source of error might be introduced that the test is designed to eliminate: the possibility of contamination by younger bones. A combined sample might, because of the nature of the faunal sequence, contain bones intrusive from later levels. The date resulting would be a rather unhelpful average, which would indicate that 'early' bones were present, but not which they were, nor their actual age. The dates obtained would also be less reliable in archaeological terms. This is where the accelerator dating method has a clear advantage. With this method, charred bones of small mass can be dated individually.

Bones were chosen from trench E, which was excavated through the full depth of the Epipalaeolithic settlement at Tell Abu Hureyra. The primary aim was to establish the age of the sheep bones by radiocarbon dating, basing this upon sufficient specimens chosen from the stratification matrix to test the Epipalaeolithic occupation, combined with specimens on bones from other large mammals where charred sheep bones were unavailable.

THE CHARRED SHEEP BONES

The samples were chosen from the Epipalaeolithic levels of Tell Abu Hureyra after the examination of all Epipalaeolithic and early Aceramic Neolithic specimens in order to see the range of variation that is present. The criteria presented by Boessneck *et al.* (1964), and Boessneck (1969) for distinguishing the bones of sheep and goat are generally satisfactory for this early material. In Table 1 below, the dating of specimens E 326 and E 286 are of particular note, in that charred seeds and charred bones from these features give very similar dates.

SPECIMEN E 285 (B224): TERMINAL PHALANX

The volar surface of the phalanx has convex margins, and does not show the narrow wedge shape found in *Capra* (Fig. 1, i, a). The margins of this surface are rounded where they merge into the medial and lateral surfaces of the phalanx, as is the dorsal margin of the phalanx, where the medial and lateral surfaces merge (Fig. 1, i, a and b). In *Capra*, the phalanx is characterised by sharp angles at these points. The extensor process is large, though the saddle-shape anterior to this is not marked (Fig. 1, i, d). The proximal end of the phalanx, viewed from its volar surface, has an oblique termination rather than the right angled form of *Capra* (Fig. 1, i, e).

SPECIMEN E 281 (B217): UNFUSED MEDIAL CONDYLE OF LEFT METATARSAL

The verticillus of the metatarsal terminates at the surfaces of the epiphysis in smooth curves (Fig. 1, ii, a). In *Capra* the verticillus teminates abruptly with an angular notch. The verticillus itself has a rounded edge, whereas in *Capra* this tends to be sharper The trochlea of the condyle is large (Fig. 1, ii, c) and the measurements of the anterior-posterior dimensions, as an index of trochlea:verticillus, is 67.0. For this dimension, Boessneck (1969) gives indices of 59.0 for the smallest specimens of *Ovis* and 62.5 for the largest specimens of *Capra*.

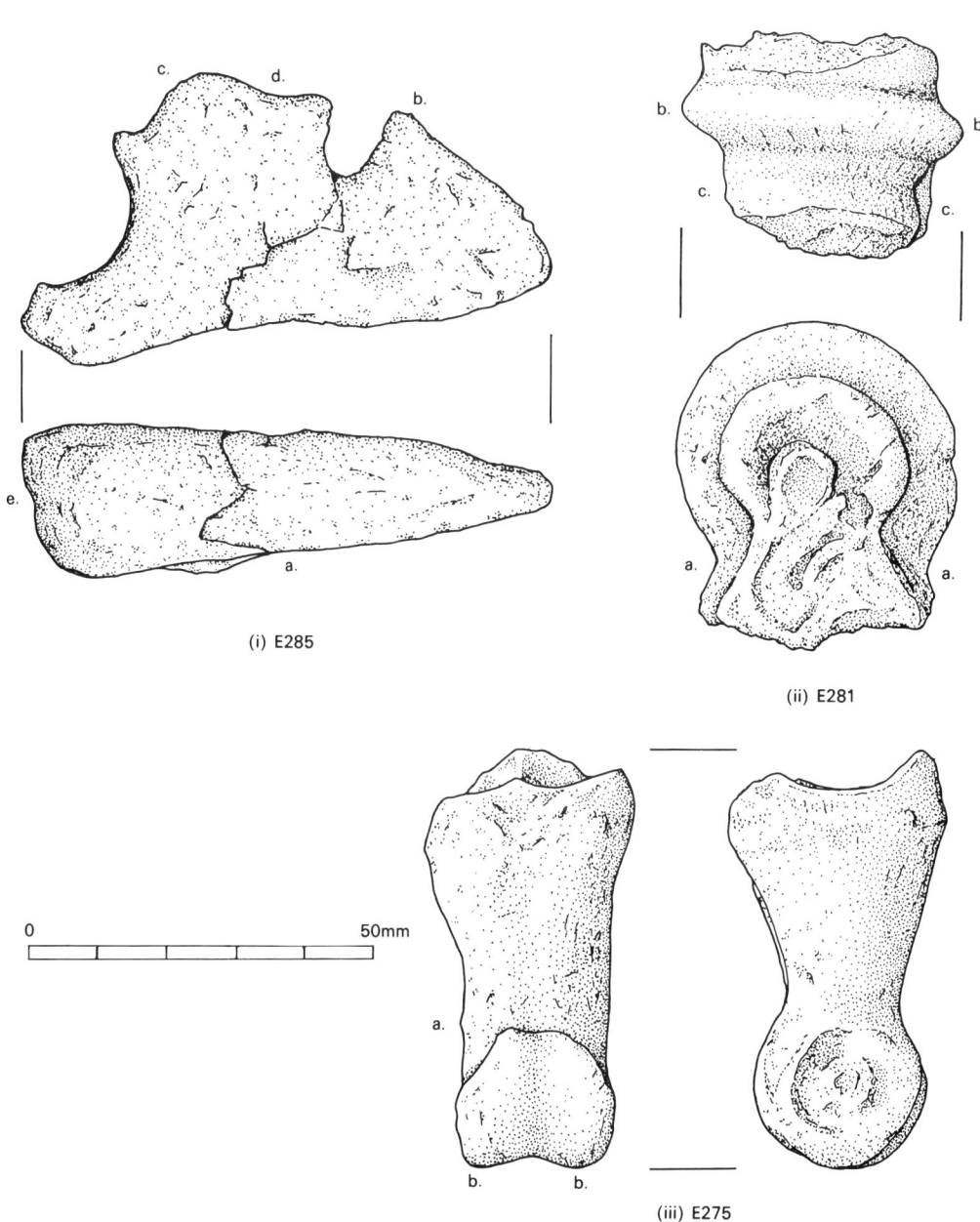

(i) E285

(ii) E281

0   50mm

(iii) E275

Fig. 2   Charred sheep bones from Tell Abu Hureyra

SPECIMEN E 275 (B217): SECOND PHALANX

On the volar surface of the phalanx the lateral margin of the distal articulation is not prolonged, nor is there evidence of this developing into a ridge as in *Capra* (Fig. 1, iii, a). The volar margin of this articulation is not indented. The two halves of the distal articulation of the phalanx, when viewed in the dorsal or volar aspects, are of about equal prominence (Fig. 1, iii, b). In *Capra* the axial part of this articulation is smaller, and the abaxial portion in consequence projects forward.

Although these features are more pronounced in the anterior phalanges than in the posterior, careful comparison of specimens from the Tell Abu Hureyra collection convinces us that the specimen illustrated is from a sheep.

The attributes of the specimens from Tell Abu Hureyra described above, and the consistent representation of these attributes within the collection, allow us to identify them as sheep with confidence.

Among the earliest features on the site, which must be close in date to its foundation, was the pit E 326. The bones from this pit did not include any specimens that could be identified as sheep with certainty. In consequence, a charred 2nd phalanx of wild cattle, *Bos primigenius*, was selected for dating. This bone was large enough for two dates to be made upon the charred material, and two upon the 'humic' content retained in the bone (see Table 1). The charred bone dates, at $11,070 \pm 160$ BP and $11,090 \pm 150$ BP are close to a charred seed date from the same pit at $10,900 \pm 200$ BP and the series shows an excellent consistency. From higher in the sequence, two charred gazelle bones give dates which fall in the correct stratigraphic order (Table 1). Features E 285, E 281 and E 275 all contained charred sheep bones which were submitted for dating after a written and photographic record was made of each. In these specimens, the dates obtained from the charred and humic fractions diverge to some degree (Table 1), with the charred material providing the older date. According to Batten *et al.* 1986, the charred fraction is likely to be the more reliable. If this proves to be so, the dates confirm the presence of sheep at Abu Hureyra at about 8000 bc and this in turn strongly indicates that the uncharred sheep bones found in older features are indeed contemporary with the sediments in which they were found.

DISTRIBUTION OF SHEEP REMAINS IN THE NEAR EAST BEFORE 9000 BP

Many of the sites described below have long occupations which continue through many thousands of years. They are presented in a broad chronological order, depending upon the first finds of sheep at the site in question.

MIDDLE PALAEOLITHIC SHEEP

Sheep bones are known from four archaeological sites with Middle Palaeolithic industries which, though not dated by radiocarbon, are likely to be older than 35,000 years in age. In Iraq, the cave of Shanidar has 10 sheep bones from the Middle Palaeolithic levels in a long and deeply statified sequence (Perkins 1964a, 1964b; Evins 1982), among a rather sparse fauna mainly of *Capra*. Perkins (1964a, 1964b) also reported 27 sheep bones from the Baradostian levels (Upper Palaeolithic) at the same site. Sheep were also found in the 'proto-Neolithic' and Aceramic Neolithic (see below).

Table 1: Accelerator dates on charred bones from Tell Abu Hureyra

Lab No.	Context	Species	Date BP	Material
OxA-387	E 326	Bos	11,070 ± 160	charred bone
OxA-468	E 326	Bos	11,090 ± 150	charred bone
OxA-469	E 326	Bos	10,920 ± 140	humic extract 1
OxA-470	E 326	Bos	10,820 ± 160	humic extract 2
OxA-172	E 326	Triticum	10,900 ± 200	charred seed
OxA-430	E 316	Gazella	11,020 ± 150	charred bone
OxA-431	E 316	Gazella	10,680 ± 150	humic extract
OxA-434	E 286	Gazella	10,434 ± 150	charred bone
OxA-435	E 286	Gazella	10,450 ± 180	humic extract
OxA-397	E 286	Triticum	10,420 ± 140	charred seed
OxA-474	E 285	Ovis	10,930 ± 150	humic extract
OxA-473	E 281	Ovis	10,000 ± 170	charred bone
OxA-472	E 281	Ovis	10,750 ± 170	humic extract
OxA-407	E 275	Ovis	10,050 ± 180	charred bone
OxA-408	E 275	Ovis	10,250 ± 160	humic extract 1
OxA-471	E 275	Ovis	10,620 ± 150	humic extract 2
OxA-475	E 252	Gazella	9060 ± 140	charred bone

At the cave of Bisitun in Iran (Coon 1951), sheep were not reported in the original faunal analysis. Recently this material has been re-examined by H-P. Uerpmann, and sheep were reported as a minor component of the fauna (Uerpmann 1978, 1979), which otherwise is mainly of equids and cervids (probably *Cervus elaphus*).

About 25 km to the east of Bisitun, is the Warwasi rock shelter (Braidwood *et al.* 1961) which, though not radiocarbon dated, has an archaeological sequence which begins in the Middle Palaeolithic and continues through the Upper Palaeolithic Baradostian to the Epipalaeolithic Zarzian culture (see below). Four sheep bones are reported from the Middle Palaeolithic levels, which otherwise have a moderate quantity of *Equus hemionus*, and the presence of *Bos primigenius* and *Cervus elaphus*. Gazelles are apparently absent (Turnbull 1975).

In Syria, Payne (1983) has studied the mammal bones from the second season of excavation at the Douara Cave and has reported sheep from the Middle Palaeolithic levels. Four bones were identified to sheep with certainty, and a further 8 were referred to this species. The associated large mammals are equids, *Camelus sp.*, *Gazella* and possible *Capra*.

UPPER PALAEOLITHIC AND EPIPALAEOLITHIC SHEEP

Within the Upper Palaeolithic of Iraq, the Palegawra Cave has sheep bones in a greater abundance, with 134 specimens reported (Turnbull and Reed 1974). The finds are from the terminal Upper Palaeolithic Zarzian culture, with radiocarbon dates of 14,400 ± 760 years, and 13,350 ± 460 years. The major elements in the fauna of Palegawra are the onager *Equus hemionus* and red deer *Cervus elaphus*. Gazelles are comparatively uncommon.

At the Warwasi Shelter, 1 sheep bone is reported from the Baradostian levels, while from the Zarzian there are only four bones identified as sheep or goat (Turnbull 1975). As noted above, no radiocarbon dates are available from the Warwasi Shelter but at the Shanidar Cave, some 400 km to the north-west, the Baradostian layer C is dated between

26,500 ± 1500 years (L-335H) and 33,630 ± 400 years (GrN-1494), with a further date of > 34,000 years (W-180). Dates for the Zarzian Culture at the caves of Shanidar and Palegawra fall between 10,600 ± 300 years (W-667) and 14,400 ± 760 years (UCLA-1703A). Goats are equally rare at Warwasi, and the mammal remains are mainly composed of equid teeth, the comparative durability of which suggests that the sparse fauna is, at least in part, the product of poor preservation.

In the north of Iran, sheep bones were reported by Setzer (in Coon 1951) from the Belt Cave, and by Frazer, Ulmer and Coon from the Hotu Cave (Coon 1952). These sites, close to Beshar in the Elburz Mountains of the south Caspian Sea, come from the region now showing the maximum genetic diversity in sheep (Nadler *et al.* 1971, 1972, 1973). The dating of the Belt Cave is uncertain, at least where the lower levels are concerned. A date of 10,560 ± 610 BP for level 11 (C-524) was accepted by Coon as an appropriate date for his 'late Mesolithic', though a date from the deeper levels 15 and 16 for the 'Mixed Mesolithic' is younger than that above, and so too is a date for levels 26 and 28, in the 'early Mesolithic'. It would be customary now to designate such industries in the Near East as 'Epipalaeolithic'. The date for level 11 appears to be broadly correct, and has been accepted as such in more recent work at the nearby cave of Ali Tappeh (McBurney 1968). At Belt Cave, sheep and goats were originally reported in all levels from the pottery Neolithic at the surface, down to level 16 in the late Mesolithic.

Dr H-P. Uerpmann (Uerpmann and Frey 1981) has re-examined the mammal remains from the Belt Cave (Gar-e Khamarband) and he has kindly confirmed to us that no wild goat bones are present, Uerpmann has thus revised the faunal identifications both in terms of the species represented and by additional material that was not included in the original report. Thus the figures as published by Coon (1951) must be disregarded. At the Belt Cave (Uerpmann 1981) domestic sheep and goat are present in the Neolithic from level 11 to the surface, while below this only wild sheep are found as a minor part of a fauna in which gazelles (*Gazella subgutturosa*) predominate. In levels 10–15, 23 bones of wild sheep are identified, of which 15 are horn core fragments. Below this, single specimens of sheep come from level 19 and the basal level 27. As the gazelles are also marked by a very high proportion of horn cores, it would seem that a marked selection was practised either during the original deposition of the bones, or in the specimens retained for study from the excavation. Thus the proportions of the mammals at the Belt Cave are probably unreliable for any numerical comparison.

The Hotu Cave (Coon 1952) contained some 7 metres of Aceramic Neolithic and later sediments, which overlay at least 5 metres of late Pleistocene sands and gravels. Both flint (Dupree, in Coon *op. cit.* 250) and bone finds were rather few. Radiocarbon dates run in a good sequence, from 11,860 ± 840 BP (P-39) in the early Mesolithic, to 9190 ± 590 BP (P-12) for the later Mesolithic and to 8070 ± 500 BP (P-37) for the Aceramic Neolithic levels. Both sheep and goat were reported from the aceramic and pottery Neolithic levels as "apparently present as domestic animals throughout". Mammal bones are rather sparse in the late Pleistocene levels at Hotu Cave, and only 203 bones of the large mammals were identified to species level. Of these an average of 56% were identified as sheep, taken to be wild. This high proportion is rather exceptional for a late Pleistocene site, and bearing in mind the revision of the Belt Cave fauna discussed above, further consideration is best suspended until this material has been re-examined.

More recent excavations at Beshar were made by McBurney (1968) in the cave of Ali

Tappeh, near to the Belt and Hotu Caves. According to McBurney's interpretation of the radiocarbon dates from the site, the lower levels (1–10) were deposited between *c*. 12,500 BP and *c*. 11,900 BP, and thus pre-date the sequences in either Belt or Hotu caves. Unlike the former sites, the Ali Tappeh stratigraphy ends with an Epipalaeolithic culture at *c*. 10,800 BP, which is overlain only by historic material. Sheep/goat were reported from all layers at Ali Tappeh, but the two species were not separated in the original report (Higgs *et al.* in McBurney 1968, p. 396). The mammal bones from Ali Tappeh have also been re-examined by Dr H-P. Uerpmann who again confirms to us that all of the bones identified as sheep/goat in the original report are certainly sheep. Uerpmann (1981, Table 2, p. 151) has also found a lower number of identifiable bones from Ali Tappeh; it would appear that there are uncertainties in the original numerical treatment of the material. However, wild sheep are common at Ali Tappeh, being present thoughout the cave sequence, and amounting to 20% of the bones of the larger terrestrial mammals. Besides the sheep, gazelles and canids are also common; equids, cattle and deer are present in rather low proportions. One important feature of this investigation is the identification of the wild sheep to the species level. Uerpmann (1981, Fig. 2, 154) shows that the sheep horn cores from the caves of Ali Tappeh and Belt (Gar-e Khamarband) have the angular frontal profile of the urial (*Ovis ammon vignei*) rather than the more rounded from of the mouflon (*Ovis ammon orientalis*).

In the Levant and in the Negev Desert, Epipalaeolithic sites of the Natufian and Harifan cultures have recently been found to have sheep bones. In the highlands of the central Negev Desert, Davis *et al.* (1982) have reported sheep bones from the sites of Ramat Harif, Abu Salem and Rosh Horesha. The sites are dated to the 9th — 10th millennium bc. Davis (1985) has also found sheep bones in another Natufian site at Hatoula, near to Latrun, about 20 km west of Jerusalem. This site has levels both of the Natufian Culture, and also of the Pre-Pottery Neolithic A period (PPNA). Nine bones of sheep were found in the Natufian levels, and one bone in the levels of the PPNA. Besides the bones of sheep, the fauna is reported as being mainly of gazelles; cervids are absent. This find is interesting in that it extends the known range of wild sheep considerably to the west, and close to the classic Natufian sites of Mount Carmel such as the caves of El Wad, Nahal Oren and Kebareh; in no case have sheep bones been reported from Natufian or earlier levels at these sites (Bate 1938; Legge 1972; Saxon 1974; Garrard 1982), nor were they found at the site of Ksar 'Akil in the Lebanon (Hooijer 1961).

EARLY ACERAMIC NEOLITHIC SHEEP

In the Zagros Mountains, sheep bones are found widely in early Neolithic village sites. The site of Zawi Chemi Shanidar (Perkins 1964a) in Iraq has sheep bones which, on the basis of a high proportion of juveniles, have been taken to represent early domestic sheep. Of a sample of 48 sheep metapodials, 54% had unfused epiphyses, and thus come from young animals. The site is dated at 10,870 ± 300 BP (W-681), and the typologically very similar level B1 at the nearby Shanidar Cave is dated at 10,600 ± 300 BP (W-667).

Further to the south in the Zagros region, the site of Tepe Asiab (Bökönyi 1977) included about 10% sheep bones among those identified as from the large mammals. Radiocarbon dates run in the correct stratigraphic order from 9050 ± 300 BP (UCLA-1714F) at the base, to 8700 ± 100 BP (UCLA-1714C) closer to the surface of the site. Bökönyi notes that the sheep at Asiab are larger than those from the nearby site of Tepe Sarab, which is dated about 1000 years later, but leaves open the question of whether they are wild or domesticated. The high proportion of wild pig and red deer also found at Tepe Asiab shows that hunting played a major part in the food supply.

A similar picture to that of Tepe Asiab is presented by the site of Ganj Dareh (Hesse 1984). The radiocarbon dates from this site appear to be too old and Hesse (1984) observes that the site was occupied "perhaps as early as the 9th millennium bc to the end of the 8th millennium bc". The fauna is marked by a very high proportion of goats, though sheep are present in all levels. Hesse (1984) argues that the goats show evidence of early domestication by their age structure, in the high proportion of young males killed before their bones were fully fused, and from a concentration upon females in the adult herd. Such a pattern corresponds to that found within domestic herds and flocks. The same avenue of investigation when applied to the bones of sheep at Ganj Dareh suggests that these were wild throughout, with no evidence of controlled slaughter nor of an age-related concentration on either sex.

By the time of the occupation of Tepe Sarab (pottery Neolithic) sheep have shown a marked increase in importance, and now represent about 60% of the larger mammals (Bökönyi 1977). There is some evidence for reduced size in the Sarab sheep when compared with those from Asiab (Bökönyi 1977, Fig. 16) and this, combined with the increased abundance of sheep and the presence of hornless individuals, does argue for their domesticated status.

In south-west Iran, the site of Ali Kosh has sheep from the earliest (Bus Mordeh) phase, which has three radiocarbon dates (Hole *et al.* 1969; Hole 1977), the earlier of which at 9900 ± 200 BP (UCLA-750D) was accepted by the excavators rather than two others at 7380 ± 180 (1–1489) and 7670 ± 170 BP (1–1496). The latter two dates certainly appear to be too late for the levels in question. On the other hand, Oates (1973) argues that the oldest date puts the Bus Mordeh phase too early and suggests that the start of this phase was about 9000 BP. The sheep remains, on the presence of a hornless individual and the age structure of the population, are taken to be domesticated. In two levels of the Bus Mordeh phase at Ali Kosh, sheep and goat average 70% of the larger mammal bones identified, and, as at Tepe Sarab, this high proportion alone indicates that both species were domestic.

At Jericho, sheep have been reported from the Aceramic Neolithic (Clutton-Brock and Uerpmann 1974) and from both the PPNA and PPNB levels. The long date list for the Jericho PPNA is given in Burleigh (1981); most dates for the PPNA levels fall between 10,300 and 9,300 BP. Two sheep bones come from the PPNA levels and 12 from the later PPNB. During the PPNA, most of the larger mammals found were gazelles, while in the PPNB caprovines become common. A similar transition is evident at Tell Abu Hureyra.

Tell Mureybit is close to Abu Hureyra but is a very much smaller site. It stands on the opposite, eastern bank of the Euphrates river. The occupation of the site begins rather later than that of Tell Abu Hureyra, and ends at about 9000 BP. The faunal sequence at Tell Mureybit has a high proportion of equids and very fluctuating proportions of *Gazella* and *Bos*. Sheep are present in all of the faunal units as studied by Ducos (1978), but fall from 8% in the first phase, to 1% or less in the second to fifth phases, rising again to 7% in the sixth and final phase. Consequently, there are few sheep bones from most levels at Tell Mureybit. The marked increase in the abundance of both sheep and goat seen at Tell Abu Hureyra in the later Aceramic Neolithic is not seen at Tell Mureybit; it seems probable that the Mureybit sequence ends before this change is manifest at sites in the region.

CONCLUSIONS

The presence of sheep in the Epipalaeolithic of Tell Abu Hureyra is of particular importance for the understanding of early sheep domestication. Our understanding of the distribution of wild sheep in this region before *c.* 7000 bc is limited by the rather sparse and widely scattered material. With the specimens from Tell Abu Hureyra, the recent finds reported at Douara Cave (Payne 1983) and from the Negev Desert and at Hatoula (Davis *et al.* 1982; Davis 1985) have extended the known range of sheep from a few archaeological sites within the classic "hilly flanks"of the Zagros Mountains and the Elburz Mountains at the southern end of the Caspian Sea to more open, steppic environments to the south and west. Sheep remains are associated with arid land mammals such as gazelles and the onager (and, at Douara Cave, camels); the cervids, associated with sheep at sites within the Elburz and Zagros mountain regions, are either absent or, as at Tell Abu Hureyra, very rare.

These finds, supported by the accelerator dates from Tell Abu Hureyra, combine to show that early sheep at archaeological sites on the steppes need not represent introduced animals, wholly or even partly under human control; sheep were a widespread steppe species, even if of only moderate local abundance. Thus the early domestication of sheep may not have been restricted to Braidwood's "hilly flanks of the fertile crescent", (Braidwood and Howe 1960, p.1) but could have been accomplished over a much wider area.

Epipalaeolithic sites within this region are generally small, and there is little evidence for the permanence of settlement that has been inferred so often. Only at Tell Abu Hureyra and Jericho are extensive Epipalaeolithic deposits found at the base of huge tells. At Abu Hureyra, the evidence from plant and animal remains indicates year-round occupation, and only at Abu Hureyra is there direct evidence for the extensive use of plant foods at this time (Hillman 1975). Besides the prevailing gazelles in the Epipalaeolithic levels, there are bones of sheep, goat, cattle and pig. By the early Aceramic Neolithic, plant cultivation is established (Hillman, pers. comm.); sheep and goat numbers show a significant increase at this time, and are second in abundance to the gazelles. The moderate proportion of caprovines, when compared to the abundance of the gazelle, is the characteristic of the early Aceramic Neolithic in the Levant (Legge 1977). However, the unimportance (in numerical terms) of domestic mammals in the food economy did not limit the development of large communities where other circumstances were favourable. Until the marked increase in the number of sites during the later Aceramic Neolithic (Moore 1981; Hole 1984), hunting must have represented greater security while the husbandry of mammals was developed. In this time of rapid economic and social change, the Epipalaeolithic settlement at Abu Hureyra, based upon a diversity of plant resources from the river valley and the mammals of the steppe, was in an unusually favoured setting. The scale and chronology of the site is almost without equal during this period and the Epipalaeolithic settlement may well present, for the first time, a convincing precursor for the emergence of an agricultural system based upon the cultivation of plants and the husbandry of animals.

ACKNOWLEDGEMENTS

We are grateful to Sebastian Payne, Trinity College, Cambridge, and Hans-Peter Uerpmann, Institut für Urgeschichte der Universitat Tubingen, for their comments upon an earlier draft of this paper.

REFERENCES

Bate, D.M.A., 1937, The fauna, in *The Stone Age of Mount Carmel*, Vol.I, (D.A. Garrod and D.M.A. Bate), Oxford: Clarendon Press.

Batten, R., Gillespie, R., Gowlett, J.A.J. and Hedges, R.E.M., 1986, The AMS dating of separate fractions in archaeology, *Proc. of the 12th Int. Radiocarbon Conf., Trondheim, Norway, 1985, Radiocarbon* 28, 2A and B.

Boessneck, J., 1969, Osteological differences between sheep (*Ovis aries* Linné) and goats (*Capra hircus* Linné) in *Science in Archaeology* (eds. D.R. Brothwell and E.S. Higgs), 2nd edn., pp. 331–358, London: Thames and Hudson.

Boessneck, J., Müller, H-H. and Tiechert, M., 1964, Osteologishe Unterscheidungsmerkmale zwischen Schaf (*Ovis aries* Linné) und Zeige (*Capra hircus* Linné), *Kühn-Archiv* 78, 1–129.

Bökönyi, S., 1977, *Animal remains from the Kermanshah Valley, Iran*, Oxford: British Archaeological Reports Supplementary Series 34.

Braidwood, R.J. and Howe, B., 1960, Prehistoric investigations in Iraqi Kurdistan, *Studies in Oriental Civilisation* 31, Chicago: University of Chicago Press.

Braidwood, R.J., Howe, B. and Reed, C., 1961, The Iranian Prehistoric project, *Science* 133, 2008–2010.

Burleigh, R., 1981, Appendix C, Radiocarbon dates, in *Excavations at Jericho III*, Text (ed. T.A. Holland), pp. 501–504, London: British School of Archaeology at Jerusalem.

Clutton-Brock, J. and Uerpmann, H-P., 1974, The sheep of early Jericho, *J. Archaeol. Sci.* 1, 3, 261–274.

Childe, V.G., 1934, *New light on the most ancient East*, London.

Coon, C.S., 1951, *Cave exploration in Iran, 1949*, Philadelphia: University Museum, University of Pennsylvania.

Coon, C.S., 1952, Excavations in Hotu Cave, Iran, 1951, a preliminary report, *Proc. American Philosophical Soc.* 96, 231–257.

Davis, S., 1985, A preliminary report of the fauna from Hatoula: a Natufian- Khiamian (PPNA) site near Latroun, Israel, in *Le Site Natoufien-Khiamien de Hatoula, près de Latroun, Israël*, (M. Lechevallier and A. Ronen), CRJF Jerusalem, Annexe B, 71–118.

Davis, S., Goring-Morris, N. and Gopher, A., 1982, Sheep bones from the Negev Epipalaeolithic, *Paléorient* 8, 1, 87–93.

Ducos, P., 1978, *Tell Mureybit: étude archéozoologique et problèmes d'écologie humaine*, Paris: Editions du Centre National de Recherche Scientifique.

Evins, M.A., 1982, The fauna from Shanidar Cave: Mousterian wild goat exploitation in northeastern Iran, *Paléorient* 8, 1, 37–58.

Garrard, A.N., 1982, The environmental implications of a re-analysis of the large mammal fauna from the Wadi el-Mughara caves, Palestine, in *Palaeoclimates, palaeoenvironments and human communities in the Eastern Mediterranean region in Later Prehistory* (eds. J.L.I. Bintcliffe and W. van Zeist), pp. 165–187, Oxford: B.A.R. International Series 133.

Hedges, R.E.M. and Gowlett, J.A.J., 1986, Radiocarbon dating by accelerator mass spectrometry, *Scientific American* 254, 1, 100–107.

Hillman, G.C., 1975, The plant remains, in *The excavation of Tell Abu Hureyra in Syria: a preliminary report*, *Proc. Prehist. Soc.* 41, 50–69.

Hole, F., 1977, Studies in the history of the Deh Luran Plain: the excavation of Chaga Sefid, *Memoirs of the Museum of Anthropology* 9, Ann Arbor, Michigan.

Hole, F., Flannery, K.V. and Neely, J.A., 1969, Prehistory and human ecology of the Deh Luran Plain, *Memoirs of the Museum of Anthropology* 1, Ann Arbor, Michigan.

Hooijer, D.A., 1961, The fossil vertebrates of Ksar 'Akil, a Palaeolithic rock shelter in Lebanon, *Zoologische Verhandelingen 49, 1–67*.

Legge, A.J., 1972, The fauna, in *Recent excavations at Nahal Oren, Israel*, (T. Noy, A.J. Legge and E.S. Higgs), *Proc. Prehist. Soc.* 39, 75–99.

Legge, A.J., 1975, The fauna of Tell Abu Hureyra: preliminary analysis, *Proc. Prehist. Soc.* 41, 74–77.

Legge, A.J. and Rowley-Conwy, P., forthcoming, Steppe hunters and the rise of domestication, *Scientific American.*

McBurney, C.B.M., 1968, The cave of Ali Tappeh and the Epipalaeolithic in N.E. Iran, *Proc. Prehist. Soc.* 34, 385–413.

Moore, A.M.T., 1981, A four-stage sequence for the Levantine Neolithic, *c.* 8500–3750 B.C., *Bulletin of the American School of Oriental Research* 246, 1–34.

Moore, A.M.T., Hillman, G.C. and Legge, A.J., 1975, The excavation of Tell Abu Hureyra in Syria: preliminary report, *Proc. Prehist. Soc.* 41, 50–69.

Nadler, C.F., Lay, D.M. and Hassinger, J.D., 1971, Cytogenetic analyses of wild sheep populations in North Iran, *Cytogenetics* 10, 137–152.

Nadler, C.F., Korobitsina, K.V., Hoffmann, R.S. and Vorontsov, N.N., 1972, Cytogenetic differentiation, geographical distribution and domestication in palearctic sheep (*Ovis*), *Sond. aus Z. f. Säugetierkunde Bd.* 38, H.2, 109–125.

Nadler, C.F., Hoffmann, R.S. and Wolf, A., 1973, G-band patterns as chromosome markers and the interpretation of chromosome evolution in wild sheep (*Ovis*), *Experientia* 29, 117–119.

Oates, J., 1973, The background and development of early farming communities in Mesopotamia, *Proc. Prehist. Soc.* 39, 147–181.

Payne, S., 1968, The origin of domestic sheep and goats: a reconsideration in the light of the fossil evidence, *Proc. Prehist. Soc.* 34, 368–384.

Payne, S., 1983, The animal bones fom the 1974 excavations at Douara Cave, in *Palaeolithic site of Douara Cave and palaeogeography of Palmyra Basin in Syria, III* (eds. K. Hanihara and T. Akazawa), University of Tokyo Bull. 21, The University Museum.

Perkins, D., 1964a, The fauna from the prehistoric levels of Shanidar Cave and Zawi Chemi Shanidar, *Report, VI Int. Congress on the Quaternary* 2, 565–572.

Perkins, D., 1964b, Prehistoric fauna from Shanidar, Iraq, *Science* 144, 1565–1566.

Protsch, R. and Berger, R., 1973, Earliest radiocarbon dates for domesticated animals, *Science* 179, 235–239.

Saxon, E.C., 1974, The mobile herding economy of Kebarah Cave, *J. Archaeol. Sci.* 1, 1, 27–46.

Turnbull, P.F., 1975, The mammalian fauna of Warwasi Rock Shelter, west-central Iran, *Fieldiana Geology* 33, 8, 141–155.

Turnbull, P.F. and Reed, C.A., 1974, The fauna from the terminal Pleistocene of Palegawra Cave, *Fieldiana Anthropology* 63, 3, 81–146.

Uerpmann, H-P., 1978, Metrical analysis of faunal remains from the Middle East in *Approaches to faunal analysis in the Near East*, (eds. R.H. Meadow and M.A. Zeder), pp. 41–45, Harvard: Peabody Museum.

Uerpmann, H-P., 1979, *Probleme der Neolithisierung des Mittelmeerraums*, Wiesbaden.

Uerpmann, H-P. and Frey, W., 1981, Die Umgebung von Gar-e Kamarband (Belt Cave) und Gar-e 'Ali Tappe (Beh-Sahr, Mazandaran, N. Iran) heute und im Spätpleistozän, in *Beiträge zur Umweltgeschichte des Vorderen Orients* (W. Frey and H-P. Uerpmann), *Beihefte Zum Tübinger Atlas Des Vorderen Orients, Reihe A (Naturwissenschaften)* 8, Wiesbaden.

Section II
Early Human Remains

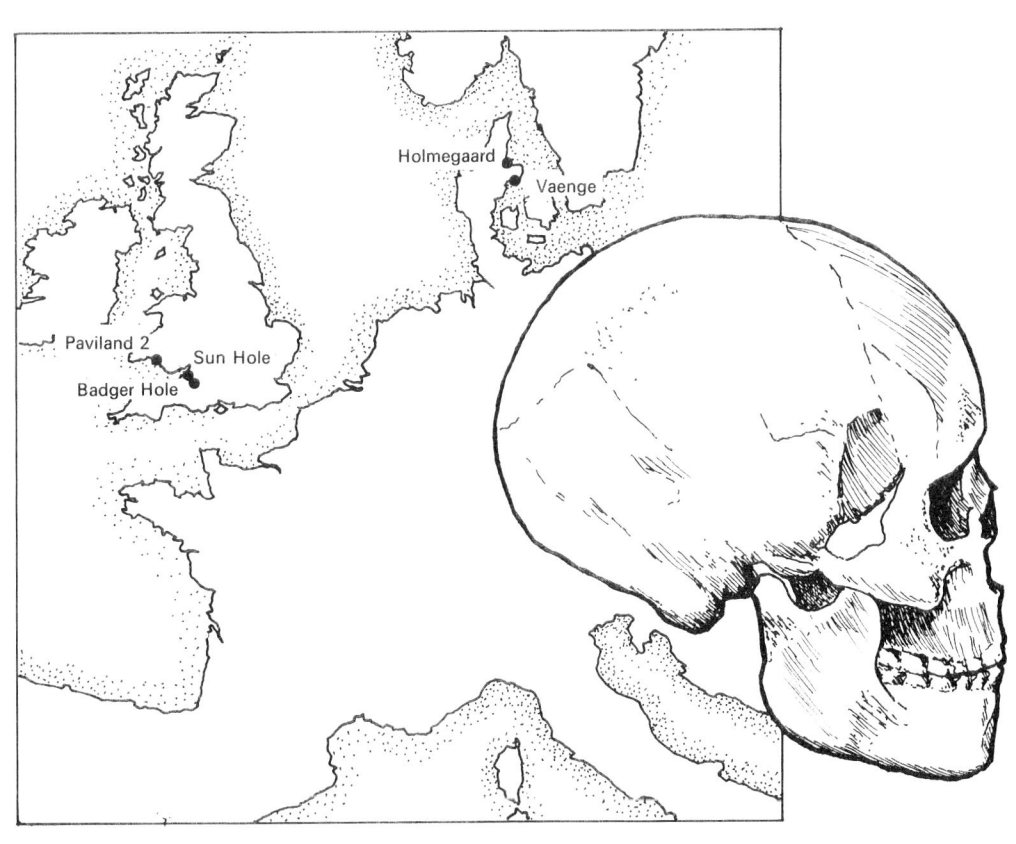

Overleaf: Some dated human remains in Europe.

NEW RADIOCARBON DATES FOR TWO MESOLITHIC BURIALS IN DENMARK

S. H. Andersen, T. S. Constandse-Westermann, R. R. Newell, R. Gillespie
J. A. J. Gowlett and R. E. M. Hedges

Skeletal remains have great intrinsic value for physical anthropologists, and can be the source of much further information than is provided by dating alone (Buikstra & Cook, 1980; Burr, 1980; Ericksen, 1976; Lengyel, 1975, 1980; Sillen and Kavanagh, 1982; Vogel and van der Merwe, 1977). Because the conventional method of radiocarbon dating requires a considerable quantity of bone (normally over 100 grams), museum directors and physical anthropologists are often hesitant to allow the dating of specimens where there is a shortage of material, especially where only single bones are available. Hence an inherent paradox: many such finds are analytically useless until they are dated but conventional dating would destroy a single-bone specimen, so that only a cast would remain for further analysis.

With the development of the radiocarbon accelerator approach it has become possible to make accurate radiometric dates from much smaller amounts of organic material, generally as little as one gram of bone (Gillespie et al. 1984; Gillespie and Gowlett, 1983). Clearly this technique of absolute dating provides a means to answers where they would not otherwise be available, by direct dating either of the skeleton, or of associated remains which would be insufficient for conventional dating.

The Oxford accelerator unit has plans for a broad programme of dating early human remains, but it seemed sensible to commence by dating Mesolithic material and to evaluate results carefully, before proceeding to Palaeolithic specimens for which the contamination problems are much greater. This is among the first papers dealing with results on human remains from Europe, though the Oxford laboratory has dated amino acid extracts from New World specimens (Bada et al. 1984) and late Pleistocene skeletons from southern Africa (Sealy and van der Merwe 1985).

The Mesolithic dating was undertaken because following the compilation of an evaluative catalogue of demonstrably and alleged Mesolithic skeletal remains in Western Europe (Newell et al. 1979), a group of finds had become apparent with a high potential for assignation to the Mesolithic period, but despite strong circumstantial evidence, the archaeological data did not provide definite resolution. Confirmation could be obtained only from direct radiometric dating of the skeletal material itself.

This paper concentrates on the material from two excavations conducted by S.H. Andersen. Two of the authors involved in the programme also expect the dating to contribute further to their inquiries in Mesolithic biology, palaeodemography, mortuary ritual and ethnicity (Constandse-Westermann and Newell, 1984a, 1984b, in press; Constandse-Westermann et al. 1985, in press; Newell and Constandse-Westermann, in press).

Among the collection of material made available through Newell and Constandse-Westermann, the Oxford laboratory commenced work, as a matter of caution, on the two most robust specimens, which came from excavations conducted by one of us (S.H.A.) in Denmark. These were a human bone, and a faunal bone associated with a burial, from the Ertebølle sites of Holmegaard-Jutland and Vaenge Sø II respectively. As it happened, samples from the same two sites had been sent to H. Tauber of the Copenhagen laboratory for C13/C12 analysis, as part of an investigation into Mesolithic diet (Tauber 1981, 1983). For these two specimens it was also possible for Tauber to carry out conventional C14 dating, whereas this would not have been possible for any of the other specimens in the Oxford series. The result was an unusual 'blind test' between two radiocarbon laboratories, operating quite different techniques, neither knowing that its results would be immediately open to comparison with another laboratory. As first published (Gillespie *et al.* 1984) the initial Oxford dates did not have a C13 correction, necessary for high accuracy here because of the very high marine component in the diet of the Mesolithic coastal population.

As a second element of comparison the Oxford samples were re-run about a year after the original dating.

(1) HOLMEGAARD-JUTLAND BURIAL

This site is at Hyllested s., Djurs sønder h., Randers amt, Jutland, Lat. 56°16′N, Long. 10°46′E.

OxA-118	Amino acids from human bone Ref. No. FHM j. nr. 1532 BA ($\delta^{13}C = -11.9‰$)	6280 ± 130 BP
OxA-533	repeat	6080 ± 80 BP
cf. K-3559	Collagen from human bone Ref. No. FHM j. nr. 1532 BA ($\delta^{13}C = -11.9‰$)	6020 ± 100 BP
K-3099	Oyster shells immediately underlying skeleton ($\delta^{13}C = +0.6‰$)	5870 ± 95 BP

Comment:
Dates on an inhumation grave, partly disturbed, with the body placed in an extended position, orientation SW-NE, found in a late Mesolithic Ertebølle coastal site (shell midden). Following the discovery, excavation of the grave and part of the site was carried out by Dr S.H. Andersen.

The skeleton belonged to a young male of 20–25 years. No grave goods were found, but two large stones covered the lower legs. The grave was clearly secondary, having been dug into the shell layer, and therefore younger than the midden, but it was impossible to obtain any clear chronological relationship between the burial and the midden.

The cultural layer of the midden was typologically homogeneous, and belongs to the older Ertebølle cultural phase, i.e. Dyrholmen I/II, most comparable with Norslund layer 2 (Andersen and Malmros 1965).

The radiocarbon dates demonstrate that the burial was indeed Mesolithic, rather than a much later intrusion, though it is somewhat older in date than might have been expected.

(2) VAENGE SØ II BURIALS

This site is at Vaenge Sø, Helgenaes s., Mols h., Randers amt, Jutland, Lat. 56°08′N, Long. 10°31′E.

OxA-117	Amino acids from wild boar cranium associated with burial Ref. No. FHM j. nr. 1850 UL + BFU ($\delta^{13}C = -19.3‰$)	5475 ± 130 BP
OxA-532	Repeat	5340 ± 80 BP
cf. K-3920	Amino acids from human bone Ref. No. FHM j. nr. 1850 BMX ($\delta^{13}C = -11.2‰$)	5500 ± 70 BP
K-3921	Amino acids from human from bone Ref. No. FHM j. nr. 1850 BMY ($\delta^{13}C = -11.1‰$)	5540 ± 65 BP

Comment:

Two graves were found on a late Mesolithic Ertebølle coastal site, lying parallel and close together on the old ground surface. They were covered by a shell midden layer, with artefacts mainly of late Ertebølle characteristics, but also with some Neolithic admixture. The skeletons were in extended positions, orientated NW-SE. There were no grave goods, but the stomach regions of both individuals were covered by two large stones. The skeletons were of one complete male (20–30 years) (BMY), and one much less complete adult female (BMX).

Because of the thin covering horizon, and the successive occupations of the site, a Mesolithic association lacked stratigraphic security. The very consistent radiocarbon dates confirm the Mesolithic date of the burials beyond question.

CONCLUSIONS

The results obtained on these and other sites show that it is possible to date human remains by the accelerator technique, using minimal amounts of material, and achieving dates closely comparable with those from an experienced conventional laboratory.

The congruence between the Oxford accelerator dates for both sites and those run concurrently by the Copenhagen radiocarbon laboratory is very encouraging. The totality of the dating confirms the Mesolithic origins and affinities of three reasonably complete extended burials.

In the case of Holmegaard-Jutland, the three dates are directly comparable, coming from the same skeleton. Although the means of the first Oxford date and the Copenhagen date differed by 260 years, this difference is not statistically significant: the pooled value is 6120 ± 80. The repeat Oxford date falls between the other two, providing a consistent group with a pooled value of 6100 ± 60.

Tauber (1981, 1983) has pointed out that dates such as these, on human skeletons where marine foods were important in the diet, do not give the true radiocarbon age even after correction for $\delta^{13}C$ fractionation: it is still necessary to take into account the reservoir effect, which gives modern marine samples from the north Atlantic an apparent age of *c.* 400 years when adjusted to the $\delta^{13}C$ standard of $-25.0‰$ PDB. As the Copenhagen laboratory takes advantage of a curious cancellation of effects, in which enrichment of $\delta^{14}C$ in carbonates and bicarbonates in the upper layers of the ocean is balanced out by depletion due to admixture of deep ocean water (Tauber 1979), the date on the oyster shells (K-3099) can be taken as giving an approximate real value in radiocarbon years. In the case of the skeletons, however, the proportion of marine diet can only be estimated. On the assumption of 50%–100% marine diet, the correction would be in the order of 200–400 years. This subtraction from the radiocarbon ages for the skeletons would bring the results into good agreement with the date for the underlying oyster shells (Tauber, pers. comm.).

In the case of Vaenge Sø it is more difficult to make strict comparisons. The two Oxford dates on the boar skull are compatible; so are the two Copenhagen dates on two separate skeletons, which were found together and have similar $\delta^{13}C$ values. As the three separate specimens may have slightly different real ages, and/or radiocarbon ages, it would be unrewarding to force the comparison further, but in general terms the agreement is satisfactory (dates for the boar, which as the $\delta^{13}C$ reading shows had no significant component of marine diet, should be rather younger than for the skeletons *if* it is exactly contemporary: there is no way to check these assumptions in detail but the dates are consistent with this).

Accelerator dates at present constitute only a small fraction of radiocarbon dates being run. In theory, it would be preferable for skeletal remains to be dated by the technique which imposes the least damage, but it is possible to take good casts and in many cases where sufficient material remains either conventional or accelerator technique can be applied satisfactorily. For most isolated bones, however, and for almost all Palaeolithic remains, the accelerator approach provides the only possible means of dating without destruction of the entire specimen.

ACKNOWLEDGEMENTS

We thank H. Tauber for comments and discussion. Technical and research assistance was provided by A. Bowles, J. Foreman, E. Hendy, M. Humm and A. Stoker. The ^{13}C measurement for the boar was provided courtesy of R. Burleigh and K.J. Matthews.

REFERENCES

Andersen, S.H. and Malmros, C., 1965, Norslund, En kystboplads fra aeldre stenalder, *Kuml* 35–114.

Bada, J.L., Gillespie, R., Gowlett, J.A.J. and Hedges, R.E.M., 1984, Accelerator mass spectrometry radiocarbon ages of amino acid extracts from Californian palaeoindian skeletons, *Nature*, 312, 442–444.

Buikstra, J.E. and Cook, D.C., 1980, Palaeopathology: an American account, *Annual Review of Anthropology* 9, 433–470.

Burr, D.B., 1980, The relationship among physical, geometrical and mechanical properties of bone, with a note on the properties of non-human primate bone, *Yearbook of Physical Anthropology* 23, 109–146.

Constandse-Westermann, T.S. and Newell, R.R., 1984a, Human biological background of population dynamics in the Western European Mesolithic, *Proc. of the Koniniklijke Nederlandse Akademie van Wetenschappen* Series B, 87, 139–223.

Constandse-Westermann, T.S. and Newell, R.R., 1984b, Mesolithic trauma: demographical and chronological trends in Western Europe, *Proc. of the 4th European Meeting of the Palaeopathology Association, Middelburg/Antwerpen 1982*, 70–76.

Constandse-Westermann, T.S., Blok, M.L. and Newell, R.R., 1985, Long bone length and stature in the Western European Mesolithic. I. Methodological problems and solutions, *Journal of Human Evolution* 14, 399–410.

Constandse-Westermann, T.S., Blok, M.L. and Newell, R.R., in press, Long bone length and stature in the Western European Mesolithic. II. Chronological and geographical variability.

Constandse-Westermann, T.S. and Newell, R.R., in press, Social and biological aspects of the Western European Mesolithic population structure: a comparison with the demography of North American Indians, *Proc. of 3rd International Symp. on The Mesolithic in Europe, Edinburgh, 1985*.

Ericksen, M.F., 1976, Cortical bone loss with age in three Native American populations, *American Journal of Physical Anthropology* 45, 443–452.

Gillespie, R. and Gowlett, J.A.J., 1983, Archaeological sampling for the new generation of radiocarbon techniques, *Oxford Journal of Archaeology* 2 (3), 379–382.

Gillespie, R., Gowlett, J.A.J., Hall, E.T. and Hedges, R.E.M., 1984, Radiocarbon measurement by accelerator mass spectrometry: an early selection of dates, *Archaeometry* 26, 1, 15–20.

Gillespie, R., Gowlett, J.A.J., Hall, E.T., Hedges, R.E.M. and Perry, C., 1985, Radiocarbon dates from the Oxford AMS system: Archaeometry datelist 2, *Archaeometry* 27, 2, 237–246.

Lengyel, I.A., 1975, *Paleoserology: blood typing with the fluorescent antibody method*, Budapest: Academiai Kiado.

Lengyel, I.A., 1980, Aging in the past: biochemical aspects of skeletal aging in recent as well as in archaeological periods, *Anthropolopiai Kozlemenyek* 42, 137–151.

Newell, R.R., Constandse-Westermann, T.S. and Meiklejohn, C., 1979, The skeletal remains of Mesolithic man in Western Europe: an evaluative catalogue, *Journal of Human Evolution* 8, 1–228.

Newell, R.R. and Constandse-Westermann, T.S., in press, Testing an ethnographic analogue of Mesolithic social structure and the archaeological resolution of Mesolithic ethnic groups and breeding populations.

Sealy, J.C. and van der Merwe, N.J., 1985, Isotope assessment of Holocene human diets in the southwestern Cape, South Africa, *Nature* 315, 138–140.

Sillen, A. and Kavanagh, M., 1982, Strontium and paleodietary research: a review, *Yearbook of Physical Anthropology* 25, 67–90.

Tauber, H., 1979, C-14 activity of arctic marine mammals, in *Proc. of the 9th International Radiocarbon Conference, Los Angeles and La Jolla, 1976*, University of California Press, 447–452.

Tauber, H., 1981, ^{13}C evidence for dietary habits of prehistoric man in Denmark, *Nature* 292, 332–333.

Tauber, H., 1983, ^{14}C dating of human beings in relation to dietary habits, in *Proc. of the 1st International Symp. on ^{14}C and Archaeology, Groningen, 1981, (eds. W.G. Mook and H.T. Waterbolk)*, PACT 8, 365–375.

Vogel, J.C. and van der Merwe, N.J., 1977, Isotopic evidence for early maize cultivation in New York State, *American Antiquity* 42, 238–242.

Ward, K.G. and Wilson, S.R., 1978, Procedures for comparing and combining radiocarbon age determinations: a critique, *Archaeometry* 20, 1, 19–31.

DIRECT DATES FOR THE FOSSIL HOMINID RECORD

C. B. Stringer

In 1981, Richard Burleigh and I published a paper which reviewed the prospects for direct dating of Upper Pleistocene, and in particular Neanderthal, fossil hominid remains (Stringer and Burleigh 1981). Our expectations of the capabilities of the accelerator radiocarbon dating technique were, in retrospect, rather optimistic, given the present practical limitations of the method, but even if earlier Upper Pleistocene fossil hominids prove to be beyond the capability of the technique, the likelihood that significant results can be anticipated in the range 30,000–50,000 years BP means that a major new approach is available for resolving one of the most intractable problems of hominid evolution – that of the origin of the living varieties of our own species *Homo sapiens*. In addition to securely placing isolated specimens within this time range, it should also be possible to resolve some long standing controversies about the antiquity of certain modern-looking fossils, which some workers believe may be intrusive or wrongly associated with earlier Upper Pleistocene deposits.

The Oxford Accelerator Unit has already provided some very interesting results for North America (Bada *et al.* 1984) and another area where results are already available is that of the British Upper Pleistocene hominid record. The lists of possible Pleistocene hominids provided in Oakley *et al.* (1971) and Campbell (1977) are extensive, but anyone who is familiar with the record will be aware of the problems in dealing with old or poorly controlled excavations, and with a failure to recognise the disturbing effects of geological processes and of living organisms (human or otherwise). A classic example of such problems is the Mendip site of Badger Hole, near Wookey Hole, in Somerset. Excavations by Balch between 1938 and 1953 produced three hominid specimens supposedly found in a breccia and apparently associated with Upper Pleistocene fauna and Upper Palaeolithic (arguably *early* Upper Palaeolithic) artefacts. However, the excavation was not properly controlled (explosives were used at times) nor published, although Balch's diaries do provide locational data on the finds which allowed Campbell (1977) to reconstruct their horizontal distribution and, less securely, their vertical distribution. McBurney (1958, published 1961) and Campbell (1968, published 1970 and 1977) conducted further small excavations, but found no other human remains and little additional archaeological material. However, Campbell collected a sample of burnt bone fragments which he was able to relate stratigraphically to a unifacial leaf point found by Balch. An 'infinite' radiocarbon age estimate of > 18,000 BP was obtained (BM-497) and this was then used as an age estimate for the three hominid specimens found by Balch (Oakley *et al.* 1971; Campbell 1977), supported by relative dating of the Badger Hole hominids (Oakley 1980). None of the three hominid fragments are robustly constructed, consisting as they do of a child's jaw (BH1 – aged *c.* 9 years), a younger child's jaw (BH2 – aged *c.* 4 years) and some cranial fragments, also thin (BH3 – attributed to an adult in Oakley *et al.* 1971). The possible presence of such gracile specimens in an early Upper Palaeolithic context was an important consideration in discussions concerning the evolution of modern humans in Europe and their arrival in Britain, but doubts surrounding their presumed antiquity have greatly hindered an assessment of their significance (Stringer *et al.* 1984).

However, in 1985 two small samples of bone were drilled from Badger Hole 1 and 3 for accelerator radiocarbon dating, and the results have dramatically clarified the situation regarding these hominid specimens (Table 1). Badger Hole 1 is clearly unrelated to Palaeolithic occupation of the site, but is of a respectable Mesolithic age, close to the conventional radiocarbon determination on the Gough's Cave 1 skeleton found about 10 km away (Barker *et al.* 1971). Sadly, one of the cranial fragments attributed to Badger Hole 3 is even more recent in age, which raises further doubts about the reliability of the relative dating techniques which led Oakley (1980) to associate these specimens with each other and with Pleistocene fauna from the site. However, before we close the file on the supposed Pleistocene hominids of Badger Hole, we should reserve judgment on the antiquity of the remaining Badger Hole 2 mandible until it too can be directly dated.

Another British site with more established credentials for genuine Pleistocene fossil hominids, although not without its dating problems, is Paviland Cave in Gower, South Wales. Here, in 1822, Buckland and others uncovered a human skeleton (the 'Red Lady', thus called because of the associated red ochre and an erroneous assessment of its sex). This specimen has been dated by conventional radiocarbon techniques to *c.* 18,500 BP (Barker *et al.* 1969) but it is difficult to relate this age determination, falling within the severe Dimlington glacial episode (Rose 1985), to the early Upper Palaeolithic occupation of the site and more generally to patterns of human settlement in western Europe (Molleson 1976; Campbell 1977; Jacobi 1980). The argument that there was a much earlier phase of Upper Palaeolithic occupation at the site seemed to be confirmed by a further conventional radiocarbon date of *c.* 27,500 BP on a bovine humerus recovered from Sollas's 1912 excavations at the site (Molleson and Burleigh 1978), and it was these excavations which produced the additional fossil hominid evidence recognised as Paviland 2 (Sollas 1913; Oakley *et al.* 1971). The metatarsal and left humerus have been regarded as of of comparable antiquity to the Paviland 1 skeleton (Oakley *et al.* 1971) but since they were recovered in excavations which produced additional Aurignacian archaeological material, the possibility remained that they were more ancient than the Paviland 1 burial. An absolute date for the Paviland 2 specimens was expected to considerably clarify the situation at the site by either closely matching the previous determination for the Paviland 1 burial, or by falling within the earlier time range expected for an Aurignacian occupation. In the event, the accelerator determination on the Paviland 2 humerus has done neither of these things (Table 1), and has further complicated the picture, since the dating evidence now points to an additional and previously unsuspected Mesolithic presence at the site! However, this new evidence may in turn lead to further reassessments of the archaeological material from Paviland.

With accelerator radiocarbon determinations on the Badger Hole and Paviland 2 hominids effectively removing three specimens from the register of putative British Pleistocene hominids, it is gratifying to also note that one dubious Pleistocene specimen has had its age confirmed by direct dating. The Sun Hole 2 ulna (not a radius as reported in Oakley *et al.* 1971) was recovered at this cave in Cheddar Gorge, Somerset, during excavations directed by Tratman between 1926 and 1928. Late Upper Palaeolithic artefacts

Table 1

Specimen	Expected age (BP)	Lab. No.	Determined age (BP)
Badger Hole 1 mandible	< 18,000	OxA-679	9060 ± 130
Badger Hole 3 cranial fragment	< 18,000	OxA-680	1380 ± 70
Paviland 2 humerus	< 18,000	OxA-681	7190 ± 80
Sun Hole 2 ulna	*c.* 12,400	OxA-535	12,210 ± 160

were apparently found in the same (5th foot) level and Campbell's additional excavations in 1968 recovered a bear humerus, in a small undisturbed area at an equivalent depth, which was dated by conventional radiocarbon techniques at about 12,400 BP (Barker *et al.* 1971; Campbell 1977). However, Jacobi questioned the relevance of this determination to human occupation at the site since he considered that archaeological material may have derived from a higher level (Jacobi 1980), so the direct determination of the age on a human bone from the site has now provided concrete evidence of a human presence at Sun Hole close to the previously determined age (Table 1). This human presence at Sun Hole may well have been related to contemporaneous human occupation at Gough's Cave on the other side of the Gorge (Jacobi, this volume).

Such direct determinations on hominid specimens illustrate the potential of accelerator radiocarbon dating in resolving long standing difficulties concerning the age of specific hominid specimens — indeed in some cases it is difficult to envisage any other approach yielding useful results. Such clarifications are of great importance for workers studying specific sites, and attempting to reconstruct the sparse evidence of early human occupation in particular areas. Once they are well dated, distinct specimens could be drawn together to complement one another, allowing palaeoanthropologists to talk about genuine population samples from particular time levels, instead of the vague or optimistic groupings of possibly disparate specimens which necessarily constitute such samples at the present time.

But beyond the problems already being tackled lies a much more significant and intractable problem, one where the radiocarbon accelerator dating technique could make an unprecendented contribution to resolving a dispute which has already lasted for over a century. This dispute concerns the origin of anatomically modern human populations in Eurasia, and their relationship to the Neanderthals. Palaeoanthropologists are as divided as ever over the question of whether *H. sapiens* evolved locally and gradually from more archaic predecessors in various parts of the world (e.g. the Neanderthals in Europe, Solo Man in Java) or whether all living peoples are derived from only one founder population in one area (probably Africa on present evidence). It has been known for many years that the earliest fossils of modern type in Europe probably date within the reliable time range of conventional radiocarbon dating, but it is only recently that the Saint-Césaire discovery of a Neanderthal partial skeleton associated with the French Upper Palaeolithic Châtelperronian industry has demonstrated that the last Neanderthals also fall within this time range (Lévêque and Vandermeersch 1981).

Here is a chance to document in detail the origins of modern humans in Eurasia. This was a major event in human evolution, the last such event, and one which is likely to remain uniquely accessible for this type of project. There is sufficient doubt at the moment about the degree of temporal and geographic overlap at the interface between Neanderthals and modern humans to allow many different scenarios for the origin of modern populations, involving varying degrees of local evolution, coexistence, hybridisation, population replacement, etc. Direct dating of critical specimens in the probable time range 30,000-40,000 BP in Europe and 40,000–50,000 BP in South West Asia should at last answer some of the great unknowns in Upper Pleistocene human evolution. If samples are available and preservation is suitable, European specimens of early modern type such as Engis, Cro-Magnon and Stetten should prove datable, as well as more controversial specimens such as Les Cottés (possibly associated with the Aurignacian) and Hahnöfersand (interpreted as a Neanderthal-modern hybrid). In eastern Europe, the robust Mladeč specimens, claimed to show 'transitional' features, and the Vindija fossils, some of which are claimed to be associated with Aurignacian artefacts, are key specimens in the debate about the origin of modern humans (Smith 1984), and should form part of any dating programme.

In south west Asia the dating problems and disagreements are, if anything, even more profound than those in Europe (Trinkaus 1984). Early 'modern' specimens fom Qafzeh (Israel) are estimated to date from over 70,000 years according to some workers, or from less than 40,000 years according to others. Such wide ranging estimates obviously make a great deal of difference to the evolutionary significance accorded to the large Qafzeh sample (e.g. as to whether they might represent ancestors for the European 'Cro-Magnon' populations). Another significant but problematic specimen from the area is the Amud 1 Neanderthal skeleton, dated by conventional radiocarbon on supposedly associated animal bones to the very young age of less than 20,000 BP and by fission track dating on animal bone to only about 28,000 BP (Suzuki and Takai 1971; Stringer and Burleigh 1981). Whether accelerator radiocarbon dating is able to resolve these and other persistent chronological problems of the area remains to be seen, but any results would be of tremendous value to palaeoanthropologists, and again it is difficult to envisage any other approach achieving even a reliable relative ordering of the specimens at present.

Beyond Europe and western Asia there are still numerous dating problems in the Upper Pleistocene, although hominid samples are generally smaller and more scattered. Linked with the possibly exotic origins of modern peoples in Europe and western Asia is the possibility that anatomically modern populations existed in Africa at an even earlier date. Sites such as Omo-Kibish (Ethiopia) and Border Cave and Klasies River Mouth Caves (South Africa) are thought by some workers to contain evidence of essentially modern-looking hominid fossils dated at more than 60,000 BP (Bräuer 1984; Rightmire 1984). If this is so, such specimens could represent the original African founder populations, from which all living peoples ultimately trace their ancestry. If their proposed dating is correct, these specimens are unlikely to be reliably dated by radiocarbon, but sufficient doubt exists in some quarters about the age of the specimens to make direct dating attempts worth considering, since as we have seen already, the technique can expose wrongly interpreted or intrusive specimens. Much more doubt surrounds the fossil hominids from sites such as Karungu, Kabua, Kanjera (Kenya) and Cape Flats (South Africa), while outside Africa the modern-looking Niah skeleton from Borneo badly needs to be substantiated as the oldest modern human specimen from eastern Asia, with its claimed age of about 40,000 BP (Oakley *et al.* 1975). Such specimens are prime candidates for direct dating to determine whether they are highly significant records of the origin and spread of modern humans, or intrusive or wrongly associated.

Clearly the success of the accelerator programme in directly dating fossil hominids depends on the availability of appropriate samples, and the suitability of samples submitted for the technique. Museum curators and palaeoanthropologists have so far shown a gratifying enthusiasm for the programme which means that there should be no shortage of material for inclusion, and as useful results accumulate, direct accelerator radiocarbon dating may well become a standard procedure in the study of Pleistocene fossil hominids of the last 50,000 years. By helping to document the origin and spread of modern humans, the accelerator programme has the potential to provide an unprecendented advance in our knowledge of the most recent events in human evolution.

REFERENCES

Bada, J.L., Gillespie, R., Gowlett, J.A.J. and Hedges, R.E.M., 1984, Accelerator mass spectrometry radiocarbon ages of amino acid extracts from California palaeoindian skeletons, *Nature* 312, 442–444.

Barker, H., Burleigh, R. and Meeks, N., 1969, British Museum Natural Radiocarbon Measurements VI, *Radiocarbon* 11, 278–294.

Barker, H., Burleigh, R. and Meeks, N., 1971, British Museum Natural Radiocarbon Measurements VII, *Radiocarbon* 13, 157–188.

Bräuer, G., 1984, A craniological approach to the origin of anatomically modern *H. sapiens* in Africa and implications for the appearance of modern Europeans, in *The origins of modern humans* (eds. F.H. Smith and F. Spencer), pp. 327–410, New York: Alan Liss.

Campbell, J.B., 1970, The Upper Palaeolithic period, in *The Mendip Hills in Prehistoric and Roman times, Bristol Archaeol. Res. Group Spec. Publ.* 1, 5–11.

Campbell, J.B., 1977, *The Upper Palaeolithic of Britain: a study of man and nature in the late Ice Age*, Oxford: Clarendon Press.

Jacobi, R.M., 1980, The Upper Palaeolithic of Britain with special reference to Wales, in *Culture and environment in Prehistoric Wales, B.A.R. British Series* 76, 15–100.

Lévêque, F. and Vandermeersch, B., 1981, Le néandertalien de Saint-Césaire, *Recherche* 12, 242–244.

McBurney, C.B.M., 1961, Two soundings in the Badger Hole, *Annual Report Wells Nat. Hist. Archaeol. Soc.* 71–72, 19–27.

Molleson, T.I., 1976, Remains of Pleistocene man in Paviland and Pontnewydd Caves, Wales, *Trans. British Cave Research Assoc.* 3, 112–116.

Molleson, T.I. and Burleigh, R., 1978, A new date for the earlier Upper Palaeolithic of Goat's Hole Cave, Paviland, Wales, *Antiquity* 52, 143–145.

Oakley, K.P., 1980, Relative dating of the fossil hominids of Europe, *Bull. Brit. Mus. Nat. Hist. (Geol.)* 34, 1–63.

Oakley, K.P., Campbell, B.G. and Molleson, T.I. (eds.), 1971, *Catalogue of fossil hominids, Vol. 2: Europe*, London: British Museum (Natural History).

Oakley, K.P., Campbell, B.G. and Molleson, T.I. (eds.), 1975, *Catalogue of fossil hominids, Vol. 3: Americas, Asia, Australasia*, London: British Museum (Natural History).

Rightmire, G.P., 1984, *Homo sapiens* in sub-Saharan Africa, in *The origins of modern humans* (eds. F.H. Smith and F. Spencer), pp. 295–325, New York: Alan Liss.

Rose, J. 1985, The Dimlington Stadial/Dimlington Chronozone: a proposal for naming the main glacial episode of the Late Devensian in Britain, *Boreas* 14, 225–230.

Smith, F.H., 1984, Fossil hominids from the Upper Pleistocene of Central Europe and the origin of modern Europeans, in *The origins of modern humans* (eds. F.H. Smith and F. Spencer), pp. 137–209, New York: Alan Liss.

Sollas, W.J., 1913, Paviland Cave, an Aurignacian station in Wales, *J. Royal Anthrop. Inst.* 43, 337–364.

Stringer, C.B. and Burleigh, R., 1981, The Neanderthal problem and the prospects for direct dating of Neanderthal remains, *Bull. Brit. Mus. Nat. Hist. (Geol.)* 35, 225–241.

Stringer, C.B., Hublin, J.J. and Vandermeersch, B.V., 1984, The origin of anatomically modern humans in western Europe, in *The origins of modern humans* (eds. F.H. Smith and F. Spencer), pp. 51–135, New York: Alan Liss.

Suzuki, H. and Takai, F. (eds.), 1970, *The Amud Man and his cave site*, Tokyo: Academic Press of Japan.

Trinkaus, E., 1984, Western Asia, in *The origins of modern humans* (eds. F.H. Smith and F. Spencer), pp. 251–293, New York: Alan Liss.

PROBLEMS IN DATING THE EARLY HUMAN SETTLEMENT OF THE AMERICAS

J. A. J. Gowlett

I. THE MAJOR ISSUES

The topic of the early colonisation of the New World stands out in prehistory as an area of the greatest and most prolonged uncertainty. Thirty years of work since the beginnings of radiocarbon has failed to answer the major questions in the minds of archaeologists to any general satisfaction. It was however a highly suitable area for operation in the early stages of accelerator dating, largely because the chronological discrepancies between the different hypotheses are so great that high precision is not required to resolve them.

There is general agreement that the Americas were first entered by a landbridge across the Bering Straits, since the emergence of modern man, setting an approximate upper time limit of 30–40,000 BP. Two broad and quite different hypotheses dominate the debate:

(1) That the first Palaeoindians were able to pass southwards only when the Canadian ice sheets split, leaving a corridor, about 12,000 years ago; and that subsequently there was very rapid settlement of both North and South America, with very rapid cultural diversification (e.g. Haynes 1969; Martin 1973).

(2) That the first occupation occurred before the glacial maximum, more than 20,000 years ago, perhaps initially through coastal colonisation along the Pacific; that during the glacial maximum occupation was concentrated within the intertropical limits; and that in the post-glacial period it expanded again rapidly both northwards and southwards (e.g. Fladmark 1979).

The key point, then, is to test rigorously whether there are any convincing signs of early occupation. Not surprisingly many American archaeologists refuse to be drawn *a priori* into one of the hypotheses outlined above, and simply await conclusive evidence, while discussing the more convincing early sites (e.g. Lynch 1983; Stanford 1982). To some extent parallels can be drawn with the search for Palaeolithic sites in Europe. In France and Britain there is an almost total lack of dated sites through the coldest period of the last glaciation, from 15,000–19,000 BP, virtually as far south as the Dordogne. Nobody expects the recolonisation to have come from the cold North. Outside the tropics and subtropics few sites might be expected in this interval anywhere in the world and particularly not in a vast recently colonised continent where population density would have to be low.

From this one could conclude that even to validate occupation at 13,000 BP, or at 12,000 BP in the far south of South America, is to make hypothesis (1) much harder to sustain: it would demand virtually instantaneous colonisation over a distance greater than that from Europe to Siberia, and/or settlement of the Bering area further back into the coldest period of the last glaciation.

Here we review the Oxford dating work of the past three years in the Americas. This has resolved several individual dating problems. If there is a paradox, it is that repeated success in skirmishes may not win a war, for one side or the other. The Oxford programme has concentrated on those human remains which were especially controversial, on archaeological sites in South America which appeared central to the '12,000 problem', together with a selection of other material from the United States and Canada which is relevant to the debate.

Although we can review the problem in these general terms, the next stage was and is to come down to dating individual human specimens and sites. Here the radiocarbon laboratory has limited influence over selection: we cannot choose the quality of sites which chance or prospection have uncovered; we can only select by omitting those where radiocarbon plainly cannot provide the answer. If, for example, claimed artefacts may not be artefacts, then the question is one for experts in lithic technology, rather than a dating laboratory. This problem clearly does not arise for human remains: but hardly anywhere in the world are these preserved often enough to be a good guide to the chronological distribution of man on their own — archaeological sites usually outnumber them by tens or hundreds to one.

2. EARLY HUMAN REMAINS

The California Palaeoindians

The skeletons from the California coast became notorious when amino acid racemisation dating apparently gave them ages of 40,000 to 70,000 years (Bada and Helfman 1975), although archaeological associations had previously suggested younger ages for some specimens (see e.g. Wormington 1957) and doubts about the early ages have always persisted (e.g. Aikens 1983). Most of the bones were poorly preserved and accelerator dating offered the only opportunity for direct dating. This was carried out by the Tucson and Oxford laboratories working independently (Table 1). The results all fall in the range 5000–9000 years (Bada *et al.* 1984; Taylor *et al.* 1983, 1984, 1985), bringing the specimens back into line with general expectation, and, at least at first glance, into accord with hypothesis (1). It appears that the racemisation estimates had been inflated through the use of the conventional radiocarbon date of *c.* 17,000 BP on the Laguna specimen for calibration (Berger 1975). The AMS date is *c.* 5000 BP (OxA-189). The new dates are significantly younger than those obtained by uranium series (Bischoff and Rosenbauer 1981).

Some reservations must be expressed. The low collagen dates are based on amino acid extracts, and are perhaps to be regarded as minimum estimates. The association with middens suggests a specialised coastal way of life. The find sites themselves are in coastal dunes. Tectonic movements must affect precision of estimates, but it is nevertheless at around 9000 years ago that sea level neared present day levels after being depressed by 100 metres or more during the last glacial maximum. Before *c.* 9000 BP coastally-adapted peoples would be invisible. In summary, 3000 years — or perhaps less — after hunters appear in the great plains , their descendants (according to hypothesis (1)) are to be found diving in the ocean the other side of the Rockies. Though this is feasible, we may wonder whether such a dramatic change of lifestyle would have occurred so suddenly in an almost empty continent.

Table 1: Accelerator dates for early human remains in the New World

OxA-186	La Jolla	5600 ± 400
OxA-188	Del Mar	5400 ± 120
OxA-774	Del Mar	5270 ± 100
OxA-189	Laguna	5100 ± 500
OxA-154	San Diego Site W-12	8470 ± 140
OxA-153	UCLA 1425	4950 ± 150
OxA-152	UCLA 1425	4850 ± 150
UCR-1437A	(Conventional) Sunnyvale	4390 ± 150
AA-50 (UCR-1437A)	Sunnyvale	3600 ± 600
AA-51 (UCR-1437D)	Sunnyvale	4650 ± 400
AA-52 (UCR-1437B)	Sunnyvale	4850 ± 400
OxA-187	Sunnyvale	6350 ± 400
AA-610A	La Jolla shores II	4820 ± 270
AA-610B	La Jolla shores II	5370 ± 250
AA-611	La Jolla shores II	6330 ± 250
AA-283	Yuha	1650 ± 250
AA-284	Yuha	3850 ± 250
AA-295	Yuha	2820 ± 200
(Chalk River Lab.)	Taber	3550 ± 500
OxA-773	Taber	3390 ± 90

Other human remains

Several other possibly early skeletons have been dated recently. These include Yuha in California (Stafford *et al.* 1984), which would never have been supposed to have an early date, except for radiocarbon measurements of *c.* 22,000 BP on caliche, and a uranium date of *c.* 19,000 (Bischoff *et al.* 1976);

The Taber skeleton in Alberta (Stalker 1969) is also quite young (see Andrews *et al.* 1984; date in Taylor *et al.* 1985). Taber was originally dated by the Chalk River accelerator, and now again by Oxford from an amino acid extract provided by J.L. Bada. The agreement of the two dates is encouraging since the two samples were prepared entirely separately. The Taber infant was in sediments probably over 100 ka old, and there was little real chance that it would have high antiquity. The same may also be true for other human remains in northern Canada.

The net result is that no human skeleton in the New World has been dated to more than 11,000 BP; archaeological dates, however, normally outnumber those on human remains, and the lack of old dates on skeletons does not argue for an absence of human occupation.

3. THE DATED SITES

Monte Verde

This site is in Chile near Puerto Montt, about 42 degrees south (Dillehay *et al.* 1982), and 11,000 km from the Bering land bridge; consequently it is useful to have the early date established by one laboratory independently confirmed. Two Oxford dates do this; they are:

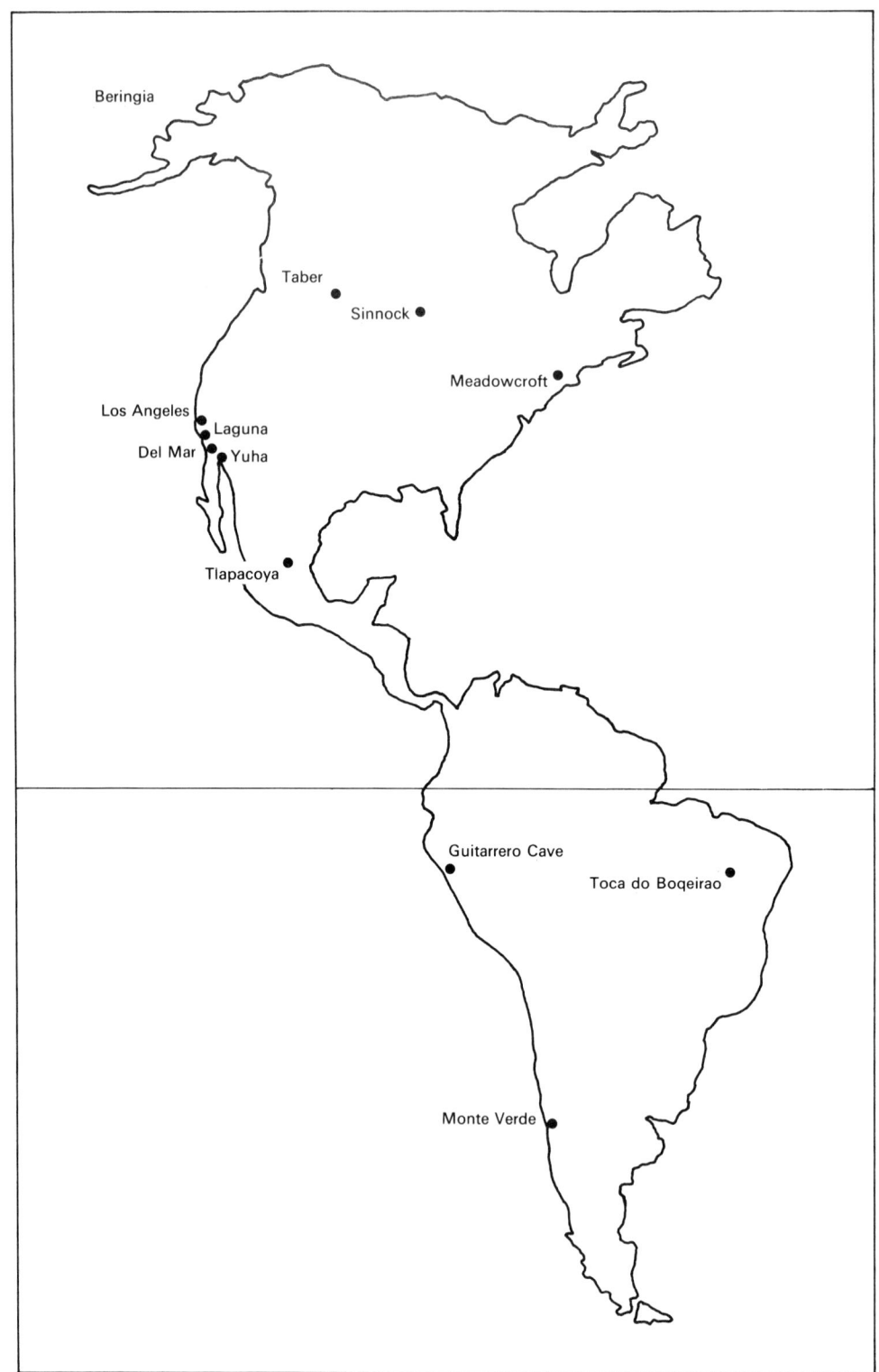

Fig. 1 Dated sites mentioned in the text

| OxA-105 Ivory Gouge | $12,000 \pm 250$ |
| OxA-381 Wood | $12,400 \pm 150$ |

These samples from the cultural levels MV-6 and the interface with MV-7 are in close agreement with the conventional dates:

| TX-3760 Bone | $11,990 \pm 250$ |
| TX-4437 Wood | $12,650 \pm 130$ |

The four dates are effectively the same. Three materials from the site thus provide similar results, and an artefact has been directly dated. The integrity of the site is thus supported, and the notion that fossil ivory was worked in later times does not have credence here. The dates are difficult to reconcile with hypothesis (1), since they would imply virtually instantaneous occupation through 80 degrees of latitude.

Guitarrero Cave

As one of the most soundly documented sites, this has offered a solid basis for the South American work (Lynch 1980; Lynch *et al.* 1985). The results on artefacts such as wooden dowels and cords confirm the 10,000 year age of these, though this had been doubted. The more controversial Geochron date of around 12,000 BP was not confirmed. The dates from that context are:

OxA-183 Charcoal, unit 62	9340 ± 150
SI-1496 Charcoal, unit 62	9475 ± 150
OxA-184 Charcoal, unit 63	9400 ± 150
GX-1859 Charcoal, unit 63	$12,560 \pm 360$
OxA-185 Charcoal, unit 64	9350 ± 150

As a very large contamination by 'dead' carbon would be necessary to make GX-1859 appear so much older, the possibility remains that it is a valid date on older charcoal (Lynch *et al.* 1985).

Meadowcroft Rockshelter

A classic case of a controversial site is provided by Meadowcroft Rockshelter in western Pennsylvania, where meticulous excavations and about 100 radiocarbon dates (Carlisle and Adovasio 1982) have not been sufficient to convince all that the human occupation goes back about 17,000 years (see Haynes 1980; Mead 1980; and reply by Adovasio *et al.* 1980). The chief argument against the early date is that the fauna is inconsistent with the glacial maximum climate, and the mechanism usually invoked to explain the discrepancy is contamination of the radiocarbon dates by 'dead' carbon contained in local geological deposits. Excellent agreement between the Oxford and Smithsonian dates (Table 2) obtained from charcoal from the base of the sequence argues against such contamination. There is no suggestion that these early dates of 30,000 are linked with human occupation, which is restricted to higher levels, but they are consistent with all other measurements. The stratigraphic consistency of the series of dates from the site would be very difficult to explain in terms of contamination, as this would be unlikely to assume a constant percentage.

Table 2: Accelerator dates for early archaeological sites in the New World

OxA-104	Guitarrero, PAn 14-102-123, wooden batten, Complex IIb	9930 ± 300
OxA-108	Guitarrero, PAn 14-102-35, wood from dowel, Complex III	10000 ± 200
OxA-109	Guitarrero, PAn 14-102-133, split dowel, Complex P	9860 ± 200
OxA-110	Guitarrero, PAn 14-102-47, wood of firedrill, Complex IV	2150 ± 150
OxA-181	Guitarrero, PAn 14-102-26, charcoal, Complex I	9520 ± 150
OxA-182	Guitarrero, PAn 14-102-60, charcoal, top of Complex I	9280 ± 150
OxA-183	Guitarrero, PAn 14-102-62, charcoal, Complex I	9340 ± 150
OxA-184	Guitarrero, PAn 14-102-63, charcoal, Complex I	9400 ± 150
OxA-185	Guitarrero, PAn 14-102-64, charcoal, base of Complex I	9350 ± 150
OxA-193	Guitarrero, PAn 14-102-18, charcoal, Complex IIe	9600 ± 130
OxA-194	Guitarrero, PAn 14-102-22, charcoal, Complex IIa	9430 ± 150
OxA-195	Guitarrero, PAn 14-102-150, wood, Complex IId	10180 ± 130
OxA-196	Guitarrero, PAn 14-102-122, cord, Complex IIc	9980 ± 120
OxA-197	Guitarrero, PAn 14-102-159, cord, Complex IIa	10340 ± 130
OxA-198	Guitarrero, PAn 14-102-82, leather	100.0 ± 1.2 (modern)
OxA-115	Sinnock, bone 5–10 cm below surface	100.0 ± 1.0 (modern)
OxA-116	Sinnock, bone c. 15 cm below surface	8030 ± 160
OxA-508	Sinnock, charcoal in pit, ? grave fill	4500 ± 80
OxA-385	Lac du Bonnet, 'elephant bone' tool	920 ± 100
OxA-363	Meadowcroft, charcoal from base of sequence	31400 ± 1200
OxA-364	Meadowcroft, humics of OxA-363	30900 ± 1100
OxA-105	Monte Verde, ivory gouge	12000 ± 250
OxA-381	Monte Verde, wood	12400 ± 150
OxA-510	Uchcumachay, Layer E-5A, bone	6670 ± 140

Sinnock and Lac du Bonnet

Although dates of less than 10,000 may seem to be of peripheral interest to the question of early settlement, this is not necessarily so. The Sinnock site lies within the area of the glacial lake Agassiz, which drained *c.* 8500 BP (Buchner 1981). A date on bone from the site was 8030 ± 160 BP (OxA-116), affirming prompt human entry into the area (Buchner in Gillespie *et al.* 1985, p. 240). It is the only early radiocarbon date within a wide radius. Here we plainly see colonisation of an area as early as local conditions would permit, at a time when occupation is already documented much further south. Since the argument holds weight here, it is difficult to dismiss it for the Great Plains where occupation appears shortly after 12,000 BP (for dates see Frison 1978.) There is little hard evidence to make colonisation from the north a more powerful argument than recolonisation from the south.

4. OUTSTANDING...

It will take accelerator dating some time to catch up with all the 'early' sites which have been found in the last forty years. In many cases dating the finds will not remove the ambiguities from the situation. Lists of radiocarbon dates on their own do little to tell us about quality of association. There can be no substitute for excavation, conducted to exacting standards, coupled with appropriate use of chronological methods. As there are a great many highly competent field archaeologists in the New World, some of the early sites will be found if they exist.

Two which would not be controversial in any other context are:

(1) Tlapacoya. Hearths excavated at Tlapacoya in the Valley of Mexico have been radiocarbon dated by the conventional technique (Mirambell 1978) with these results:

A-794	Charcoal, Tlapacoya Site 1, Alpha	24,000 ± 4000 BP
I-4449	Charcoal, Tlapacoya Site 1, Alpha	21,700 ± 500 BP

In each of these cases abundant ash and charcoal were found in a hearth overlying an ancient beach, associated with faunal remains and stone tools.

(2) Toca do Boqueirão. A hearth excavated at this site in the state of Piaui, Brazil, is alleged to have yielded the oldest parietal art yet known in South America (Guidon 1983): fallen fragments of rock wall had been used as hearth stones, and one of them, underlying the hearth, had traces of paint on it. The hearth date was:

Gif-5397 Charcoal	17,000 ± 400 BP

In a controversial atmosphere it would be standard practice to look for errors of dating or of association in these dates. Yet in general erroneous dates on wood or charcoal tend to be made too young by contamination, rather than too old. Nevertheless, on some sites such as Meadowcroft, charcoal is found which does not have human associations, and the strength of association can never be taken for granted. These conventional dates by rights ought to have solved the debate centred on hypotheses (1) and (2), had we but the confidence to accept them. As most archaeologists may be too cautious to do so, rightly pointing to many other sites which did not pass stringent tests, accelerator dating still has a considerable future in this field. Nevertheless, already, the general pattern of reliably established dates is most economically explained in terms of a first occupation beginning before the glacial maximum of 18,000 BP, but not necessarily long before it. Recent work on linguistic diversity of native American languages (Rogers 1985) appears to support this view independently. Dates from Toca do Boqueirão published since this paper was written (Guidon and Delibrias 1986) point yet more strongly to such early occupation.

ACKNOWLEDGEMENTS

We thank J.M. Adovasio, J.L. Bada, A.P. Buchner, T.D. Dillehay and T.F. Lynch for their assistance in providing samples and other co-operation.

REFERENCES

Adovasio, J.M., Gunn, J.D., Donahue, J., Stuckenrath, R., Guilday, J.E. and Volman, K., 1980, Yes Virginia, it really is that old: a reply to Haynes and Mead, *American Antiquity* 45 (3), 588–595.
Aikens, C.M., 1983, The Far West, in *Ancient North Americans* (ed. J.D. Jennings), Freeman & Co., San Francisco.
Andrews, H.R., Ball, G.C., Brown, R.M., Burn, N., Davies, W.G., Imahori, Y., Milton, J.C.D. and Workman, W., 1984, Accelerator mass spectrometry at Chalk River, *3rd Int. Symposium on Accelerator Mass Spectrometry, Proc., Nuclear Instruments and Methods in Physics Research* 233 (B5), 2, 134–138.
Bada, J.L., Gillespie, R., Gowlett, J.A.J. and Hedges, R.E.M, 1984, Accelerator mass spectrometry radiocarbon ages of amino acid extracts from Californian Palaeoindian skeletons, *Nature* 312, 442–444.

Bada, J.L. and Helfman, P.M., 1975, Amino acid racemization dating of fossil bones, *World Archaeology* 7, 160–173.

Berger, R., 1975, Advances and results in radiocarbon dating: Early man in America, *World Archaeology* 7, 174–184.

Bischoff, J.L., Merriam, R., Childers, W.M. and Protsch, R., 1976, Antiquity of man in America indicated by radiometric dates on the Yuha burial site, *Nature* 261, 128–129.

Bischoff, J.L. and Rosenbauer, R.J., 1981, Uranium series dating of human skeletal remains from the Del Mar and Sunnyvale sites, California, *Science* 213, 1003–1005.

Buchner, A.P., 1981, Sinnock: a Paleolithic camp and kill site in Manitoba, *Papers in Manitoba Archaeology, Final Report No.10.*

Carlisle, R.C. and Adovasio, J.M., 1982, *Collected papers on the archaeology of Meadowcroft Rockshelter and the Cross Creek Drainage*, 7th Annual Meeting of the Society for American Archaeology, Minneapolis. Pittsburg.

Dillehay, T.D., Pino, Q.M., Davis, E.M., Valastro, S. Jr., Varela, A.G. and Casamiquela, R., 1982, Monte Verde: radiocarbon dates from an Early-Man site in south-central Chile, *Journal of Field Archaeology* 9, 547–550.

Fladmark, K.R., 1979, Routes: alternate migration corridors for early man in North America, *American Antiquity* 44, 55–69.

Frison, G.C., 1978, *Prehistoric hunters of the High Plains*, New York, Academic Press.

Gillespie, R., Gowlett, J.A.J., Hall, E.T. and Hedges, R.E.M., 1985, Radiocarbon dates from the Oxford AMS system: Archaeometry Datelist 2, *Archaeometry* 27 (2), 237–246.

Guidon, N., 1983, Contribution a l'étude de l'art rupestre de l'Amérique du Sud, *L'Anthropologie* 87, 2, 257–270.

Guidon, N. and Delibrias, G., 1986, Carbon-14 dates point to man in the Americas 32,000 years ago, *Nature* 321, 769–771.

Haynes, C.V., 1969, The earliest Americans, *Science* 166, 709–715.

Haynes, C.V., 1980, Paleoindian charcoal from Meadowcroft Rockshelter: is contamination a problem? *American Antiquity* 45 (3), 582–587.

Lynch, T.F. (ed.), 1980, *Guitarrero Cave: Early Man in the Andes*, New York, Academic Press.

Lynch, T.F. 1983, The Paleo-Indians, in *Ancient South Americans* (ed. J.D. Jennings), pp. 87–137, San Francisco: Freeman.

Lynch, T.F., Gillespie, R., Gowlett, J.A.J. and Hedges, R.E.M., 1985, Chronology of Guitarrero Cave, Peru, *Science* 229, 864–867.

Martin, P.S., 1973, The discovery of America, *Science* 179, 969–974.

Mead, J.I., 1980, Is it really that old? A comment about the Meadowcroft 'Overview', *American Antiquity* 45 (3), 579–582.

Mirambell, L., 1978, Tlapacoya: a late Pleistocene site in central Mexico, in *Early man in America from a circum-Pacific perspective* (ed. A.L. Bryan), pp. 221–230, Occasional Paper No. 1 of the Department of Anthropology, University of Alberta, Edmonton, Alberta.

Rogers, R.A., 1985, Glacial geography and native North American languages, *Quaternary Research* 23, 1, 130–137.

Stafford, T.W., Jr., Jull, A.J.T., Zabel, T.H., Donahue, D.J., Duhamel, R.C., Brendel, K., Haynes, C.V. and Taylor, R.E., 1984, Holocene age of the Yuha burial: direct radiocarbon determinations by accelerator mass spectrometry, *Nature* 308, 446–447.

Stalker, A. MacS., 1969, Geology and age of the early man site at Taber, Alberta, *American Antiquity* 34, 425–428.

Stanford, D.J., 1982, A critical review of archaeological evidence relating to the antiquity of human occupation of the New World, in *Plains Indians studies* (eds. D.H. Ubelaker and H.J. Viola), pp. 202–218, *Smithsonian Contributions to Archaeology*, No. 30, Washington.

Taylor, R.E., Payen, L.A., Gerow, B., Donahue, D.J., Zabel, T.H., Jull, A.J.T. and Damon, P.E., 1983, Middle Holocene age of the Sunnyvale skeleton, *Science* 220, 1271–1273.

Taylor, R.E., Payen, L.A. and Slota, P.J. Jr., 1984, Impact of AMS ^{14}C determinations on considerations of the antiquity of *Homo sapiens* in the western hemisphere, *3rd Int. Symposium on Accelerator Mass Spectrometry, Nuclear Instruments and Methods in Physics Research* 233 (B5), 2, 312–316.

Taylor, R.E., Payen, L.A, Prior, C.A., Slota, P.J., Jr., Gillespie, R., Gowlett, J.A.J., Hedges, R.E.M., Jull, A.J.T., Zabel, T.H., Donahue, D.J., Stafford, T.W. and Berger, R., 1985, Major revisions in the Pleistocene age assignments for North American human skeletons: none older than 11,000 ^{14}C years B.P., *American Antiquity* 50 (1), 136–140.

Wormington, H.M., 1957, *Ancient Man in North America*, Denver Museum of Natural Popular Series, No. 4, Denver.

Section III
The Upper Palaeolithic

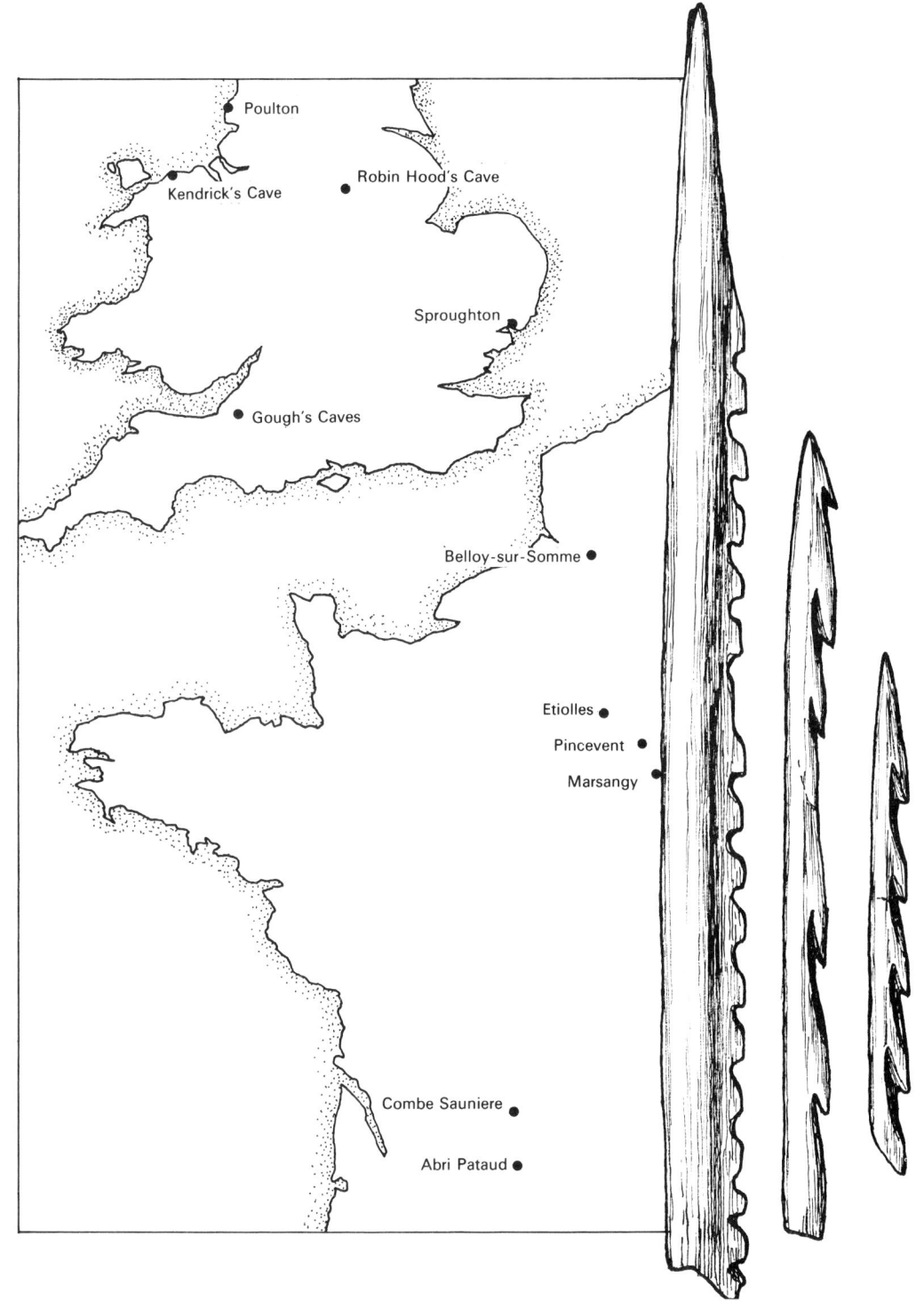

Poulton

Kendrick's Cave

Robin Hood's Cave

Sproughton

Gough's Caves

Belloy-sur-Somme

Etiolles

Pincevent

Marsangy

Combe Sauniere

Abri Pataud

Overleaf: Accelerator-dated Upper Palaeolithic sites in Britain and France; (left) barbed bone point from Sproughton (date OxA-517), after Wymer et al. 1975; (right) barbed bone points dated by association with the Poulton elk, after Campbell (1977).

LESSONS OF CONTEXT AND CONTAMINATION IN DATING THE UPPER PALAEOLITHIC

J. A. J. Gowlett and R. E. M. Hedges

I. THE AREA OF OPERATION

From the beginning of Accelerator Dating it has been appreciated that one of the major archaeological contributions would be to the Palaeolithic. Discussion papers on this topic were presented both at Oxford and Tucson even before the machines were capable of dating. Now the plans have become a reality, but as yet we are only part way along the line. The main emphasis of the discussion papers was on extension of the radiocarbon range. In the meantime it has become more and more evident that there is a great deal for accelerator dating to sort out in those periods of the Upper Palaeolithic which were already within the time range of conventional dating in principle, but only accessible in favourable circumstances of sampling.

By 1984 it was apparent that background problems and contamination difficulties would make any advance to older dates (over 40,000 years) very hard going. There were both scientific and archaeological reasons for concentrating on the later periods.

Some of the fruits of that first phase are reviewed in the following papers. Once it was known that approved projects would deal with the earlier Upper Palaeolithic in France (Mellars and Bricker below), and with the Upper Palaeolithic in Britain (Jacobi, this volume; Cook and Barton, this volume), it was possible to map out a suitable area of activity for the laboratory in its experimental work. This was aimed to bridge the gap between the other projects both geographically and temporally, and thus to strengthen as much as possible the pattern of comparisons which could be made.

The rationale for this was as follows:
(1) We had to assess whether very small samples could really be used for dating Palaeolithic sites effectively, given the questions of context which were bound to arise.
(2) We needed to compare the results of conventional and accelerator dating in as many circumstances as possible.
(3) As the main areas of National Facility operation were after and before the maximum of the last glaciation (i.e. 10–13 ka on the one hand, and 23–30 ka on the other), and as the preservation and contamination problems are very different, it was highly desirable to have a series of dates running right through the whole period.

The resulting dates are thus concentrated on important open sites in northern France, which can be compared with sites in Britain, and on rockshelter sequences further south in France which run from the late Palaeolithic back to and through the last glacial maximum (dates listed in Gillespie et al. 1985 and Gowlett et al. 1986; summarised in Gowlett 1986). Much of the experimental work has however centred on sample chemistry (e.g. Gillespie et al. 1984b), and thus it ranges through the geographical areas, including Britain.

Table 1: Dating Comparisons for Upper Palaeolithic time range

Guitarrero Cave

OxA-194	IIa, Unit 22 charcoal	9430 ± 150
SI-1499	IIa, Unit 22 charcoal	9580 ± 135
OxA-197	IIa, Unit 159 cord	$10,340 \pm 130$
SI-1502	IIa, Unit 159 charcoal	$10,240 \pm 110$
OxA-181	I, Unit 26 charcoal	9520 ± 150
SI-1498	I, Unit 26 charcoal	9660 ± 150
OxA-183	I, Unit 62 charcoal	9340 ± 150
SI-1496	I, Unit 62 charcoal	9475 ± 130

Monte Verde, MV-6/MV-7

OxA-105	Ivory gouge	$12,000 \pm 250$
TX-3760	Bone	$11,990 \pm 250$
OxA-381	Wood	$12,400 \pm 150$
TX-4437	Wood	$12,650 \pm 130$

Pincevent, Level IV2 (upper silts IVa)

OxA-148	Bone, 27 L89	$12,600 \pm 200$
OxA-467	Charcoal, 37 M121	$12,250 \pm 160$
Gif 6283	Charcoal, 27 M89	$12,120 \pm 130$

Pincevent, Level IV213 (lower silts IVb)

OxA-149	Bone, 25 K81	$12,400 \pm 200$
OxA-176	Bone, 25 M79 26	$12,000 \pm 220$
OxA-177	Bone, 25 M80	$12,300 \pm 220$
Gif 6284	Charcoal, 26 K88-89	$11,800 \pm 130$

Gough's Cave

BM-2183	Atlas of Equus ferus (collagen)	$12,120 \pm 120$
OxA-589	Collagen of BM-2183	$12,340 \pm 150$
OxA-590	Amino acids from collagen of BM-2183	$12,370 \pm 150$
BM-2187	Metapodial of Equus ferus (collagen)	$12,070 \pm 170$
OxA-591	Collagen of BM-2187	$12,260 \pm 160$
OxA-592	Amino acids from collagen of BM-2187	$12,500 \pm 160$

Wadi Kubbaniya

OxA-103	Charcoal, E-78-3	$17,150 \pm 300$
AA-96	Charcoal, E-78-3	$17,450 \pm 1000$
Range of SMU dates:		
SMU-591	Charcoal	$16,960 \pm 210$
SMU-599	Charcoal	$18,240 \pm 290$

Krakow-Spadzista Street

OxA-635	Ivory, Site C2-4	$20,200 \pm 350$
Ly-2542	Bone, Site C2	$21,000 \pm 900$
GrN-6636	Charred bone, Site B	$23,040 \pm 170$
Ly-631	Bone, Site B	$20,600 \pm 1050$

Abri Pataud, Middle Perigordian VI, Layer 3, Lens 2a

| OxA-163 | Whole collagen | $23,180 \pm 670$ |
| OxA-164 | Amino acids of same sample | $24,250 \pm 750$ |

OxA-165	Amino acids of same sample	$24,440 \pm 740$
GrN-4506	(bone from same bag)	$22,780 \pm 140$
GrN-4721	(bone from same bag)	$23,010 \pm 170$

Meadowcroft Rockshelter

SI-1687	Charcoal	$30,710 \pm 1140$
OxA-363	Charcoal residue	$31,400 \pm 1200$
OxA-364	Humics of OxA-363	$30,900 \pm 1100$

2. BASIC ACCURACY ASSESSED THROUGH COMPARISONS

A problem in all radiocarbon dating beyond the Holocene is the lack of known age material, for which the true date is precisely known by independent means, and hence the absence of any calibration curve, even in the most general terms. Other techniques such as TL and Uranium series cannot offer high-precision comparisons, and therefore all radiocarbon daters are compelled to work in radiocarbon years, knowing that contamination by more recent material is the main obstacle to reaching dates that are true (accurate) radiocarbon ages. The principal external yardstick for comparison therefore comes from other laboratories, especially those which have specialised in dating beyond 30,000 years (notably Groningen). Any contamination introduced during processing in the laboratory can be best assessed through the measurement of backgrounds (see e.g. Gillespie and Hedges 1984).

Such comparisons and intercomparisons break new ground, as in the past all formal intercomparisons have dealt with recent material. To underline the relevance of this, we can consider the effect of one per cent contamination by modern carbon at various ages: at 2000 years the effect would be undetectable; at 10,000 years it would take 200 years off the date; at 28,000 years it would increase the ^{14}C activity by over 30 per cent, and reduce the date to less than 26,000 years. A real age of 37,000 radiocarbon years would be reduced to about 32,000.

This suggests that the great majority of existing dates over 20,000 years (apparent) may be underestimates. Operators of a new technique cannot therefore instantly go for or assume absolute success, but must measure themselves against the best available. Even this is difficult on archaeological sites, since there is also the question of obtaining suitable sample material. Old bone dates are rare, because conventional laboratories require large quantities of material. Nevertheless bone is in principle an ideal material for dating, since the amino acids which make up collagen can be highly purified (Gillespie *et al.* 1984b). One reservation will be discussed below: where the collagen is almost entirely degraded (less than 5% surviving), amino acids or other persistent contamination from other sources may bias the result to some degree.

Several intercomparisons with other laboratories are available from charcoal. Most of these are outside Europe, where charcoal is quite rare on Pleistocene sites, but they extend through almost the entire Upper Palaeolithic time range. In general these comparisons show very close agreement with conventional laboratories (Table 1):

Abu Hureyra: Here the Oxford dates on charred grain agree well with BM-1718 ($11,160 \pm 110$ BP), BM-1723 ($10,700 \pm 500$ BP) and BM-1121 ($10,792 \pm 92$) on collected charcoal (Burleigh *et al.* 1982a and b; see also Legge and Rowley-Conwy, this volume).

Meadowcroft: A 'blind' comparison with the Smithsonian Laboratory yielded very good agreement at 30,000 years (Gowlett, this volume).

Wadi Kubbaniya: A date on charcoal matched those from the Texas laboratory and the Arizona accelerator (Gillespie *et al.* 1984a, p.17).

Guitarrero: Dates on charcoal at c. 9–10 ka generally show good agreement with the Smithsonian laboratory (Lynch *et al.* 1985; Gowlett *et al.* 1986).

Pincevent: The Oxford charcoal samples are in broad agreement with those from the Gif Laboratory, though not all from precisely the same contexts (Gowlett *et al.* 1986, p.120).

These good agreements demonstrate the nonsense of the assertion, occasionally heard, that accelerator dates will be different from conventional dates simply because it is a different technique of measurement. Charcoal is not however in principle the best material for any kind of radiocarbon dating: there is no means of guaranteeing or assaying the purity of the prepared sample. We cannot be sure that the charcoal dates are all as old as they ought to be, merely that the preparation systems of the various laboratories attain a similar measure of success in eliminating contamination.

Few bone/charcoal intercomparisons are available for Pleistocene material, largely because charcoal is rarely well-preserved on Palaeolithic sites in Europe (and most of it is bone charcoal), while bone is usually too poorly preserved for dating in the Middle East and Africa. Pincevent (Leroi-Gourhan and Brézillon 1972) provides an example where bone and charcoal dates are in good agreement (Table 2).

Bone is inevitably the mainstay of Palaeolithic dating in Europe. Purification of amino acids is much more feasible on the small scale required for accelerator dating.

The following comparisons on bone are particularly relevant:

(a) BM Gough's Cave. The intercomparison with British Museum dates on bone from Gough's Cave (see Burleigh, Jacobi, below) is especially useful for tying down the scale of possible discrepancies in the late Palaeolithic. The Oxford dates are 300–400 years older on average than the BM dates. For two samples a highly controlled intercomparison was arranged, in which collagen purified by the BM laboratory was dated at Oxford, in two forms: (i) combusted without further treatment, and (ii) converted to amino acids. It appears that about 100 years of the difference in dates between laboratories can be ascribed to the conversion from collagen to amino acids (Gowlett *et al.* 1986, p.118).

(b) Monte Verde, Chile. An Oxford date on mastodon ivory agrees closely with a Texas date on bone from the same level (Gillespie *et al.* 1985).

(c) Abri Pataud. The dates from this site are reviewed in Mellars and Bricker (this volume). One particular comparison again shows general comparability with a conventional laboratory (Table 1). In this case it should be noted that the Groningen date was on a different bone from the same context; the Oxford amino acid date is apparently older than the collagen date, but the difference is not statistically significant.

(d) Northampton Rhinoceros. This specimen has been dated on several occasions (Burleigh *et al.* 1984; Gillespie *et al.* 1984b). The age for the specimen was distinctly older when a more highly purified sample was used: viz., first total amino acids, rather than collagen; then specific single amino acids common in collagen:

BM-2074	Initial collagen preparation	$23,880 \pm 770$
OxA-98	Accelerator, total amino acids	$26,300 \pm 500$
BM-2074C	Conventional total amino acids	$25,500 \pm 650$
OxA-155	Accelerator, proline	$28,800 \pm 1100$
OxA-156	Accelerator, hydroxyproline	$28,000 \pm 1100$

(e)   Krakow. The date of $20,200 \pm 350$ on mammoth bone from a dwelling at Krakow-Spadzista Street C2 (OxA-635) is entirely compatible with Lyon and Groningen dates from a slightly lower level (Kozlowski and Kubiak 1972).

The general implication of these results on bone is that there is again usually good agreement between accelerator and conventional dating, but with a tendency for the Oxford dates to come out older, particularly for older samples. Although the pattern requires much further documentation it tends to reinforce the idea that most existing bone dates over 20,000 BP may represent considerable underestimates of the true radiocarbon ages.

(f)   Aveley. An extreme case of the potential for field contamination is provided by the Aveley elephant. This specimen of *E. antiquus* must be at least 100,000 years old, and has almost no remaining collagen. Extraction of amino acids from large quantities of bone (around 100g) gave the following results:

OxA-370 Aveley Elephant	$20,640 \pm 630$
OxA-623 Repeat, separate sample	$21,100 \pm 400$

The extracted amino acids thus have an activity of *c.* 6% modern.

3. CAN QUESTIONS OF CONTEXTUAL RELIABILITY AND SAMPLE CONTAMINATION BE DISENTANGLED (?)

Intercomparisons where the samples are different pieces of bone or charcoal would not form a good pattern, such as those listed above, unless the contexts in question were relatively undisturbed. But already by the time 50 accelerator samples had been dated, it was evident that numbers of very small samples differed in date from their expected contexts. Usually these were grains or small animal bones. To what extent could such effects bias or jeopardise the effect of small sample dating for the Palaeolithic ?

This is a crucial question, because there are unfortunately indications that in circumstances of very poor bone preservation, rare in Europe but all too common in the Middle East and Africa, even total amino acid dates give an underestimate of the real age.

The following case studies give some measure for assessing this problem:

Belloy-sur-Somme. Several dates show that this site recently excavated by J.-P. Fagnart (1984) has an age of *c.* 10,000 BP. But one horse tooth gave an age of *c.* 8000 BP (OxA-461), which would have been the first occurrence of the species in this period in northern France. On investigation, the identification was confirmed, but we noted that the amino acid yield had been very low indeed. The specimen was redated from a part which gave a higher yield, and the resulting date was $10,110 \pm 130$ (OxA-722), fully in accordance with the others from the site (Table 2).

Etiolles. Early on in the programme five bone specimens were dated from a single level at Etiolles (Taborin *et al.* 1979; Barbetti *et al.* 1980). Four results were in close agreement, but the fourth apparently a thousand years younger (Gowlett *et al.* 1986). This was subsequently seen to be the specimen with lowest collagen content.

Marsangy. Three specimens were dated from one locality at Marsangy (Schmider 1979; Delporte *et al.* 1982). Two produced dates in the range 11.6–12.2 ka, but the third on a reindeer bone with low collagen content was $9,770 \pm 180$ (OxA-505). B. Schmider notes that there is no indication that the specimen is intrusive. Reindeer has however been dated to this period elsewhere (e.g. Clutton-Brock and Burleigh 1983).

Pincevent. The sites at Pincevent are relatively deeply buried in fine-grained sediments with laminations (Leroi-Gourhan and Brézillon 1972). There are few signs of disturbance. The bone dates from the site have been consistent in the range 11.9–12.6 ka (apart from younger results on a charred bone, which may be intrusive).

Combe Saunière. A series of well-preserved specimens was provided by J.-M. Geneste from his recent excavations of the Combe-Saunière sequence (see contribution by J.-M. Geneste, pp. 409–410 in Rigaud 1982). These run through from Late Magdalenian to Perigordian. In a general sense the dates reflect the stratigraphic sequence of the site, and its expected time range, but there are several inversions, and in two cases bones found side by side produced dates over 10,000 years apart. Indications of Palaeolithic digging had been found on the site, and M. Geneste stresses the complexity of processes which could have acted in the formation of this site.

At least some generalisations can be made from these data. There are specific documented instances, as at Belloy, or Aveley, where bone has hardly any of the original collagen remaining, and appears to contain contaminating amino acids. There are cave sites and talus sites where the material is clearly somewhat disturbed and mixed, and where the pattern of discrepancies cannot be explained by any known form of contamination, but only by movement of bones or charcoal upwards or downwards. There are many cases where this is well documented for well-preserved samples. On other sites we can only work by inference at the moment, since collagen levels are extremely low, and it is not yet possible to determine conclusively which factors provide the explanation. As the 'young' date at Marsangy is compatible with other dates for reindeer, it cannot be dismissed as a date, though the older dates fit the lithic industry better. At Etiolles, it may be best to give the archaeological context the benefit of the doubt, and to assume that the 'youngest' sample of the five suffers from contamination. The relative order of TL dates established by Valladas (1981) for these sites fits the broad pattern of the radiocarbon dates.

So far the only general conclusion which can be drawn is that a pattern of determinations will nearly always effectively date a site, even though a proportion of individual samples may have been mobile within the site, or present contamination problems. Naturally one wishes to know certainly in all cases which samples are intrusive, and which, if any, are giving false radiocarbon readings through contamination.

On the basis of all our bone dates, including known age material, and the examples above, we can summarise:
(1)   that bone dates on well preserved collagen have proved highly reliable;
(2)   that problematic dates have only occurred in samples where collagen content is less

Table 2. *Upper Palaeolithic dates with amino acid values*
expressed as milligrams per gram bone

OxA-138	Etiolles, Foyer N20, N21,157	12,990 ± 300	37.0
OxA-139	Etiolles, Foyer N20, O20,224	13,000 ± 300	48.5
OxA-173	Etiolles, Foyer N20, O20,222	12,800 ± 220	11.4
OxA-174	Etiolles, Foyer N20, O21,332	11,900 ± 250	9.6
OxA-175	Etiolles, Foyer N20, O21,332	12,900 ± 220	40.3
OxA-148	Pincevent, Upper Level, IV27(2)	12,600 ± 200	15.0
OxA-149	Pincevent, Lower Level, IV25	12,400 ± 200	15.3
OxA-176	Pincevent, Lower Level, IV25	12,000 ± 220	19.6
OxA-177	Pincevent, Lower Level, IV25	12,300 ± 220	20.7
OxA-391	Pincevent, Level III2 27, P85	11,870 ± 130	33.0
OxA-178	Marsangy, P16, reindeer antler	11,600 ± 200	8.1
OxA-505	Marsangy, B12-35 reindeer bone	9770 ± 180	8.2
OxA-740	Marsangy, C14-85 reindeer tooth	12,120 ± 200	11.3
OxA-460	Belloy, bone, upper horizon B117	5255 ± 80	
OxA-461	Belloy, tooth, E. ferus, B117,17	8010 ± 110	
OxA-722	Belloy, repeat on B117,17	10,110 ± 130	47.4
OxA-462	Belloy, tooth, E.ferus, B117,44	9720 ± 130	1.35
OxA-723	Belloy, tooth, E.ferus, B131,H17.21	9890 ± 150	42.5
OxA-724	Belloy, tooth, E.ferus, B131,I19.9	10,260 ± 160	54.8
OxA-517	Sproughton Point 1 (30a) bone	10,910 ± 150	33.8
OxA-518	Sproughton Point 2 (30b) antler	10,700 ± 160	105.0
OxA-370	Aveley Elephant	20,640 ± 630	
OxA-623	Aveley Elephant	21,100 ± 400	0.3

than 5% of modern. Even in these cases the decrease in age is moderate, and could be accounted for by a small input of contaminating material, usually equivalent to 1–5% modern.

In a Palaeolithic context a Neolithic or later period find would always be distinguishable instantly beyond doubt.

4. GENERAL VALUE OF THE PROJECT

This project has allowed the dating of numbers of open sites and rockshelters which in one way or another were beyond the reach of conventional dating. For the first time it has allowed a network of intercomparisons to be built up between laboratories on Palaeolithic sites. A priority now is for further work on the chemistry of low collagen bone, partly so as to obtain greater confidence in evaluating existing sets of dates, and partly so as to achieve dates on more sites which are at present undatable.

ACKNOWLEDGEMENTS

The Oxford Radiocarbon Accelerator Unit is largely supported by an SERC Research Grant. We thank our colleagues R.Gillespie, A.D. Bowles, J.F. Foreman, E. Hendy, M.J. Humm, C. Perry and A.J. Stoker who have been largely responsible for this work; and all the archaeologists who have been so helpful in providing dating material.

REFERENCES

Barbetti, M., Taborin, Y., Schmider, B. and Flude, K., 1980, Archaeomagnetic results from late Pleistocene hearths at Etiolles and Marsangy, France, *Archaeometry* 22, 1, 25–46.

Burleigh, R., 1986, Complementarity of conventional and accelerator dating: examples in Pleistocene extinctions, this volume.

Burleigh, R., Matthews, K. and Ambers, J., 1982a, British Museum Natural Radiocarbon Measurements XIV, *Radiocarbon* 24, 3, 229–261.

Burleigh, R., Ambers, J. and Matthews, K., 1982b, British Museum Natural Radiocarbon Measurements XV, *Radiocarbon* 24, 3, 262–290.

Burleigh, R., Ambers, J. and Matthews, K., 1984, British Museum Natural Radiocarbon Measurements XVII, *Radiocarbon* 26, 1, 59–74.

Campbell, J.B., 1977, *The Upper Palaeolithic of Britain*, Oxford: Clarendon Press.

Clutton-Brock, J. and Burleigh, R., 1983, Some archaeological applications of the dating of animal bone by radiocarbon with particular reference to post-pleistocene extinctions, in *1st Int. Symposium on ¹⁴C and Archaeology, Groningen, Proc., 1981* (eds. W.G. Mook and H.T. Waterbolk), pp. 409–418, *PACT* 8.

Cook, J. and Barton, R.N.E., 1986, Dating Late Devensian-Early Flandrian barbed points, this volume.

Delporte, H., Mons, L. and Schmider, B., 1982, Sur un rognon de silex, en forme de statuette féminine, provenant du gisement du Pré-des-Forges à Marsangy (Yonne), *Bulletin de la Société Préhistorique Française* 79, 275–278.

Fagnart, J.-P., 1984, Le Paléolithique superieur dans le Nord de la France: un état de question, *Bulletin de la Société Préhistorique Française* 81, 291–301.

Gillespie, R. and Hedges, R.E.M., 1984, Laboratory contamination in radiocarbon accelerator mass spectrometry, *Nuclear Instruments and Methods in Physics Research* B5, 294–296.

Gillespie, R., Gowlett, J.A.J., Hall, E.T. and Hedges, R.E.M., 1984a, Radiocarbon measurement by accelerator mass spectrometry: an early selection of dates, *Archaeometry* 26, 1, 15–20.

Gillespie, R., Hedges, R.E.M. and Wand, J.O., 1984b, Radiocarbon dating of bone by accelerator mass spectrometry, *J. Archaeol. Sci* 11, 1, 165–170.

Gillespie, R., Gowlett, J.A.J., Hall, E.T., Hedges, R.E.M. and Perry, C., 1985, Radiocarbon dates from the Oxford AMS system: Archaeometry Datelist 2, *Archaeometry* 27, 2, 237–246.

Gowlett, J.A.J., 1986, Radiocarbon accelerator dating of the Upper Palaeolithic in Northwest Europe: a provisional view, in *Recent studies in the Palaeolithic of Britain and its nearest neighbours* (ed. S.N. Collcutt), Sheffield: J.R. Collis Publications, Department of Archaeology and Prehistory, Sheffield University.

Gowlett, J.A.J., Hall, E.T., Hedges, R.E.M. and Perry, C., 1986a, Radiocarbon dates from the Oxford AMS system: Archaeometry Datelist 3, *Archaeometry* 28, 1, 116–125.

Gowlett, J.A.J., Hedges, R.E.M., Law, I. and Perry, C., 1986b, Radiocarbon dates from the Oxford AMS system: Archaeometry Datelist 4, *Archaeometry* 28, 2, 206–221.

Jacobi, R.M., 1986, A.M.S. results from Cheddar Gorge — trodden and untrodden 'lifeways', this volume.

Kozlowski, J.K. and Kubiak, H., 1972, Late Palaeolithic dwellings made of mammoth bones in South Poland, *Nature* 237, 463–464.

Legge, A.J. and Rowley-Conwy, P., 1986, New radiocarbon dates for early sheep at Tell Abu Hureyra, Syria, this volume.

Leroi-Gourhan, A. and Brézillon, M., 1972, Fouilles de Pincevent: essai d'analyse ethnographique d'un habitat magdalénien, *Gallia Préhistoire suppl.*, Vol.7.

Lynch, T.F., Gillespie, R., Gowlett, J.A.J. and Hedges, R.E.M., 1985, Chronology of Guitarrero Cave, Peru, *Science* 229, 864–867.

Mellars, P. and Bricker, H.M., 1986, Problems in dating the earlier Upper Palaeolithic, this volume.

Rigaud, J.-Ph., 1982, Circonscription d'Aquitaine, *Gallia Préhistoire* 25, 407–436.

Schmider, B., 1979, Un nouveau facies du Magdalénien final du Bassin parisien: L'industrie du gisement du Pré-des-Forges, à Marsangy (Yonne), in *La Fin des Temps glaciaires en Europe* (ed. D. de Sonneville-Bordes), pp. 763–771, Colloques internationaux du CNRS No. 271, Talence, Paris: CNRS.

Taborin, Y., Olive, M. and Pigeot, N., 1979, Les habitats paléolithiques des bords de Seine: Etiolles (Essonne, France) in *La Fin des Temps glaciaires en Europe* (ed. D. de Sonneville-Bordes), pp. 773–782, Colloques internationaux du CNRS No. 271, Talence, Paris: CNRS.

Valladas, H., 1981, Datation par thermoluminescence de grés brulés de foyers de quatre gisements du Magdalénien final du Bassin Parisien, *C.R. Acad. Sc. Paris, Series II*, 292, 355–358.

RADIOCARBON ACCELERATOR DATING IN THE EARLIER UPPER PALAEOLITHIC

P. A. Mellars and H. M. Bricker

The results discussed here form part of a larger project, aimed at clarifying the chronology of the earlier stages of the Upper Palaeolithic over a range of areas in Europe and the Near East and — ultimately — at the dating of the critical interface between the Middle and Upper Palaeolithic stages in these regions. The project is in its initial stages, and the work has focused so far on the dating of a number of sites in the Perigord region of south-western France, chosen partly because of the inherent archaeological importance of the sites, and partly because of the impressive quality of the documentation of the sites in stratigraphic, archaeological and palaeoenvironmental terms.

So far a total of 34 separate samples has been dated from four sites — the Abri Pataud (16 samples), La Ferrassie (6 samples), Le Flageolet I (6 samples) and the Abri du Facteur (Tursac) (6 samples). The site of Abri Pataud in particular was chosen because of the exceptionally detailed information available on the stratigraphic provenance and associations of the samples (see Movius 1975, 1977; Bricker and David 1984; David 1985). In all, this site contains 14 major levels of Upper Palaeolithic occupation, spanning the time range from the earlier Aurignacian (*c.* 34,000 BP) to the early Solutrian (*c.* 20,000 BP) and containing rich, intermediate levels of Middle Perigordian ('Perigordian IV'), Noaillian ('Perigordian Vc'), Final Perigordian ('Perigordian VI') and 'Proto-Magdalenian' occupation. Most of these levels contain an internal sequence of finer stratigraphic divisions, and the entire sequence is now fully documented in terms of the archaeological, palaeontological and palaeoenvironmental aspects of the different occupation horizons. Perhaps the most important feature of the site from the perspective of the present project is the opportunity to compare the results of accelerator dating with the extensive series of dates produced during an earlier dating programme by the Groningen laboratory (Vogel and Waterbolk 1963, 1967, 1972; Waterbolk 1971; Movius 1977). In all, the Groningen laboratory has produced a total of 34 individual dates spanning the whole range of the Abri Pataud deposits and — with a few notable exceptions — providing a generally coherent and consistent chronology for the entire sequence. In terms of dating by conventional radiocarbon techniques, the Abri Pataud ranks as the most extensively dated Upper Palaeolithic site in Europe. In many ways, therefore, the Abri Pataud offered an ideal context in which to compare the relative performance of accelerator techniques and conventional ^{14}C procedures in dating samples within the earlier part of the Upper Palaeolithic time range — i.e., broadly from 35,000 to 20,000 BP.

The results of the accelerator dating of the Abri Pataud samples are summarized in Table 1 and compared directly with the earlier results of the the Groningen dating programme in Figure 1. All of the accelerator dates are based on individual fragments of animal bone, for which the stratigraphic and spatial locations were documented during the excavations. The

Table 1

Oxford accelerator dates for the Abri Pataud sequence. All of the dates are based on the total amino-acid fraction of animal bone samples (see Gillespie *et al.*, 1984).

Lab. No.	Level	Cultural Phase	Date BP
OxA-373	Level 1	Lower Solutrian	20,400 ± 450
OxA-162	Level 2 : Lens 2	Proto-Magdalenian	22,000 ± 600
OxA-163	Level 3 : Lens 2A	Final Perigordian ('Perigordian VI')	23,180 ± 670
OxA-164			24,250 ± 750
OxA-165			24,440 ± 740
OxA-599	Lens 3		21,740 ± 450
OxA-686			24,500 ± 600
OxA-580	Eboulis 3–4 Red	Noaillian (late)	20,400 ± 600
OxA-687			25,500 ± 700
OxA-166			26,100 ± 900
OxA-167	Level 4a		26,500 ± 980
OxA-374	Level 4 : Lens M-1		26,300 ± 900
OxA-168	Lens 0–3	Noaillian (early)	26,900 ± 1000
OxA-169	Level 5 : Lens K-1	Middle Perigordian ('Perigordian IV')	28,400 ± 1100
OxA-581	Lens R-3		26,000 ± 1000
OxA-582	Level 6 : Lens 1	Late Aurignacian	24,340 ± 700

earlier samples dated by the Groningen laboratory consisted of much larger, bulked samples of bone, in some cases collected from several square metres of the respective deposits. The accelerator measurements are based exclusively on the extracted amino acid fraction of the bones (Gillespie *et al.* 1984). The Groningen dates were based mainly on the 'whole collagen' fraction of the samples, though with extensive chemical pretreatment to remove contamination by humic acids or other potential contaminants (Vogel and Waterbolk 1967). The standard deviation values quoted for the Oxford measurements apply to the total error for the system as a whole. This includes the errors in the accelerator measurement itself and the effects (more significant at ages $> 12,000$ years) of variability in the levels of laboratory contamination during processing of the samples. (The latter is assessed on the basis of frequent cross-checks on control samples known to contain no ^{14}C.) The errors introduced by possible field contamination cannot be estimated, although the pretreatment is designed to minimise these. The relative effects of these 'background' factors of course increase sharply with the absolute age of the samples, and for samples in the range of 20–30,000 years typically yield standard deviation ranges of the order of ± 500 to ± 1000 years.

The essential comparisons between the new accelerator measurements and the earlier series of Groningen dates are shown clearly in Figure 1. The overall level of agreement between the two series of dates is clearly impressive, although it will be seen that a small number of the dates produced by both the Groningen and Oxford laboratories are quite clearly inconsistent with respect to the overall pattern of dating for the site as a whole. The problems posed by these occasional 'anomalous' dates in the Oxford and Groningen series are discussed further below, but can hardly obscure the more general consistencies apparent in the total range of dates secured for the Abri Pataud sequence as a whole. Where direct comparisons can be made between the Oxford and Groningen measurements, the results almost invariably coincide within the ranges of a single standard deviation. In the

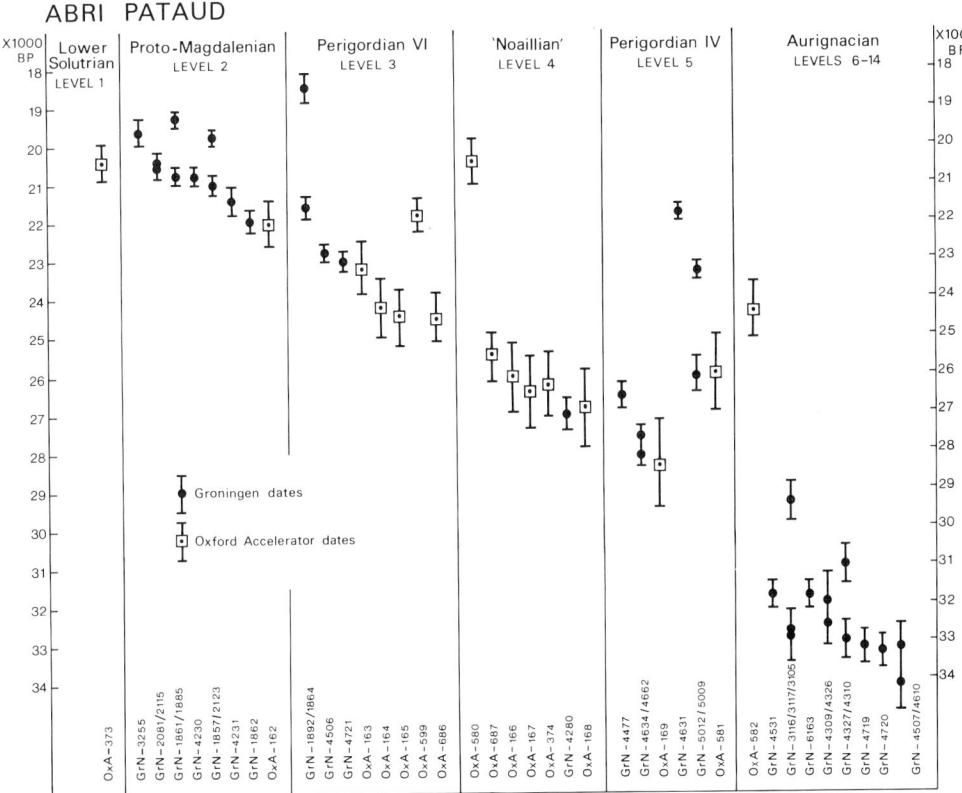

Fig. 1 Comparison between Oxford accelerator dates for the Abri Pataud sequence and the earlier series of dates by the Groningen laboratory. All of the Groningen dates are based on samples of burnt or unburnt animal bone, with the exception of GrN-3115, 3116 and 3117 which are based on samples of charcoal. For the unburnt bones, the dates are based on the extracted collagen fraction of the samples; for the burnt bones, separate measurements are shown for the soluble ('extract') and insoluble ('residue') fractions of the samples (see Vogel and Waterbolk 1963, 1967, 1972; Waterbolk 1971). All of the Oxford measurements are based on the total amino-acid fraction of unburnt bones. For details of Oxford samples, see Table 1.

cases where apparent discrepancies between the accelerator dates and the earlier Groningen measurements are discernible (for example in some of the dates for the Perigordian IV, Perigordian VI and Proto-Magdalenian levels) there is a general tendency for the accelerator dates to be slightly older than the corresponding measurements by the Groningen laboratory for the same levels. The most likely explanation of this pattern is that the selective extraction of the amino acid fraction of the bones employed for the Oxford measurements has been more effective in eliminating recent contamination from the samples than that achieved by the measurement of the whole collagen fraction applied to the Groningen samples (see Gillespie *et al.* 1984). But in general terms the close correspondence between the majority of the accelerator measurements and the earlier measurements by the Groningen laboratory is the most striking feature of the samples dated so far.

The problems posed by the four anomalous dates in the Oxford series (OxA-580, OxA-581, OxA-582 and OxA-599) are at present being investigated further in the laboratory. As noted above, similar problems in the dating of individual bone samples had already been encountered in the earlier series of measurements by the Groningen laboratory (see Figure 1; also Vogel and Waterbolk 1967; Waterbolk 1971). In both cases the 'problematic' dates show a similar pattern, in revealing occasional dates which are substantially *younger* (by amounts ranging from *c.* 2000 to as much as 5000 years) than the 'expected' age of the associated deposits as predicted from the overall stratigraphic sequence of the C-14 measurements. One initial question which must clearly be considered is the possibility of some form of stratigraphic displacement of these particular samples between different stratigraphic levels in the sequence. There are however a number of arguments that can be advanced against this possibility. In the case of the Oxford samples, the individual bone samples were deliberately selected from areas of the site where the stratigraphy of the deposits was particularly clear and well defined. For the Groningen samples the situation is even clearer, since it is impossible to visualize any form of undetected stratigraphic disturbance which could displace the large, bulked samples of animal bone employed for the radiocarbon measurements through (in some cases) between two and three metres of well stratified deposits. And of course if large-scale displacements of this kind were involved in certain levels of the site, then this should show up equally clearly in the typology of the associated artefacts from the respective occupation levels (for example, by the occurrence of Noailles burins in the Perigordian IV level, Gravette points in the Aurignacian levels and so on). Occurrences of this kind are virtually unknown in the Abri Pataud sequence. If stratigraphic displacements of this kind can be ruled out, then the most likely explanation for the occurrence of the occasional 'young' dates in both the Oxford and Groningen series would seem to lie in the presence of some residual contamination of the samples by more recent, intrusive carbon, which was not completely eliminated during the laboratory cleaning and preparation of the samples. In the case of the Groningen measurements, this has already been accepted by Vogel and Waterbolk (1967, pp. 115–116; Waterbolk 1971, pp. 17–18, 29) as the only plausible explanation for the occasional aberrant dates in the Abri Pataud series.

As Gillespie *et al.* (1984) have already emphasized, one of the crucial factors in the radiocarbon dating of very old samples is the quality of preservation of the original collagen content in the dated samples. Clearly, if contamination factors are involved, these will be largely dependent on the ratios between quantities of modern, intrusive carbon, and the surviving content of original (i.e. ancient) carbon in the respective samples. Preliminary data suggest that a clear relationship of this kind does exist, and that the problematic dates in the Abri Pataud series coincide with the samples in which the levels of surviving collagen in the bones are exceptionally low. Current work in the laboratory is therefore concentrating on the detailed chemistry of these samples, with a view to identifying any potential, residual sources of contamination in the extracted amino-acid fractions.

ARCHAEOLOGICAL IMPLICATIONS

Clearly the most important general result of the Abri Pataud measurements is the generally close correspondence between the results of the new accelerator measurements, and the earlier series of conventional ^{14}C dates produced by the Groningen laboratory. As noted earlier, where discrepancies do occur, the new accelerator measurements are generally slightly older than those for the same stratigraphic levels produced by the Groningen laboratory, suggesting that for samples with relatively high levels of preserved collagen, the selective extraction and measurement of the amino acid component of the samples is more effective in eliminating potential contaminants than the measurement of the whole collagen fraction. In terms of the detailed chronology of the Abri Pataud sequence, the new dates are important in confirming that the age of the Perigordian IV occupation levels extends back to at least 28,000 BP, that the age of the 'Noaillian' levels (previously dated by only a single radiocarbon measurement) is in the region of 26,000–27,000 BP, and that the age of the Proto-Magdalenian level probably lies between 21,000 and 22,000 BP. A rather more surprising discovery is that the age of the Final Perigordian ('Perigordian VI') occupation — previously dated to *c.* 23,000 BP — must extend back to at least 24,000 – 24,500 BP. As yet no samples have been measured by the accelerator technique for the earliest Aurignacian levels; the dates secured for the overlying Perigordian IV levels however indicate that the whole sequence of Aurignacian occupations at the Abri Pataud (comprising levels 6 to 14) had come to an end well before 28,000 BP, while the earlier Groningen measurements demonstrate that the basal levels of the Aurignacian sequence must reach back to at least 33,000 – 34,000 BP (Waterbolk 1971; Movius 1977).

Work is currently in progress in Oxford on the dating of samples from several other sites in south-west France with archaeological sequences covering broadly the same time range as the Abri Pataud sequence — including La Ferrassie, Le Flageolet I, and the Abri du Facteur (Tursac). Only preliminary results are at present available for these sites but a number of the dates obtained so far have an important bearing on the relative and absolute chronology of these sequences in relation to the Abri Pataud succession (see Table 2). Measurement of samples from the site of Le Flageolet I (Rigaud 1982) has produced dates of $24,600 \pm 700$ BP for a level with a final Upper Perigordian industry (layers I–III), $25,700 \pm 700$ BP for a level with a late Noaillian assemblage (layer V), $26,500 \pm 900$ BP for an earlier Upper Perigordian horizon with Font Robert points (layer VI) and $33,800 \pm 1800$ BP for an early Aurignacian level (layer XI). All of these dates are in close agreement with both the archaeological comparisons and detailed geological correlations between this site and the Abri Pataud sequence (see David and Bricker 1986; Laville *et al.* 1980). Dating of samples from La Ferrassie has yielded two particular useful dates of $27,530 \pm 720$ and $27,900 \pm 770$ BP for levels containing rich and typical Perigordian industries with Font Robert points (Delporte 1984). These dates are important in confirming that the age of these levels falls between the age of the Perigordian IV and Noaillian levels in the Abri Pataud succession. Unfortunately there is rather less agreement between these dates and the existing series of dates produced by the Gif-sur-Yvette laboratory for the La Ferrassie sequence (Delibrias 1984). The new accelerator dates seem in general to be substantially older than the dates for the corresponding stratigraphic levels recorded by the Gif laboratory, in several cases by between 2000 and 5000 years (see Figure 2). As Delibrias has already suggested (1984, p. 106), this may reflect the fact that many of the samples from La Ferrassie were collected from close to the edges of long-exposed sections, which may have allowed substantial contamination of the samples by modern humic materials.

Table 2

Oxford accelerator dates for samples from La Ferrassie, Le Flageolet I and Abri du Facteur (Tursac). All dates are based on the total amino-acid fraction of animal bone samples. Samples from La Ferrassie and Abri du Facteur were supplied by H. Delporte, and samples from Le Flageolet I by J-P. Rigaud. Further details of the measurements are published in Gowlett *et al.* 1986.

Lab. No.	Level	Cultural Phase	Date BP
LA FERRASSIE			
OxA-401	Level B7 (Frontal)	?Late Perigordian	23,800 ± 530
OxA-402	Level D2x (Frontal)	Perigordian V ('Font Robert')	27,900 ± 770
OxA-403	Level D2h (Frontal)		27,530 ± 720
OxA-404	Level E		26,250 ± 620
OxA-405	Level G1 (Saggittal)	Aurignacian III/IV	29,000 ± 850
OxA-409	Level K4 (Frontal)	Aurignacian II	28,600 ± 1050
LE FLAGEOLET			
OXA-448	Levels I-III	Late Perigordian	24,600 ± 700
OxA-596	Level IV	(?) Late Noaillian	23,250 ± 500
OxA-447	Level V	Late Noaillian	25,700 ± 700
OxA-579	Level VI	Upper Perigordian (Font Robert points)	26,500 ± 900
OxA-597	Level VIII-1	Late Aurignacian	24,800 ± 600
OxA-598	Level XI	Early Aurignacian	33,800 ± 1800
ABRI DU FACTEUR (TURSAC)			
OxA-583	Level 10/11	Noaillian	24,720 ± 600
OxA-584	,,	,,	24,210 ± 500
OxA-585	,,	,,	24,400 ± 600
OxA-586	,,	,,	24,690 ± 600
OxA-594	,,	,,	25,450 ± 650
OxA-595	,,	,,	25,630 ± 650

CONCLUSIONS

From the results discussed above it is clear that the current programme of accelerator dating at Oxford has already made important contributions to the understanding of earlier Upper Palaeolithic chronology in western Europe. As emphasised in the introduction, these are only the initial results of a much larger dating programme, and samples are at present being assembled from a wider range of sites, both in France and other areas of Europe and the Near East. There seems little doubt that accelerator dating has a potentially critical role to play in resolving many of the current debates and uncertainties over the relative and absolute chronology of the earlier stages of the Upper Palaeolithic and − eventually − of the critical transition between the Middle and Upper Palaeolithic stages in different regions. In this context, accelerator dating has three crucial advantages over conventional dating methods. Firstly, the capacity to date extremely small samples of material means that greatly increased numbers of samples are potentially available for dating, so that multiple, closely stratified samples can be dated from specific sites or specific archaeological horizons as a direct check on the internal coherence and consistency of the dates obtained. Secondly, the ability to date *individual* fragments of bone (as opposed to large, bulked samples) allows the *precise* stratigraphic position of each dated sample to be documented accurately − both

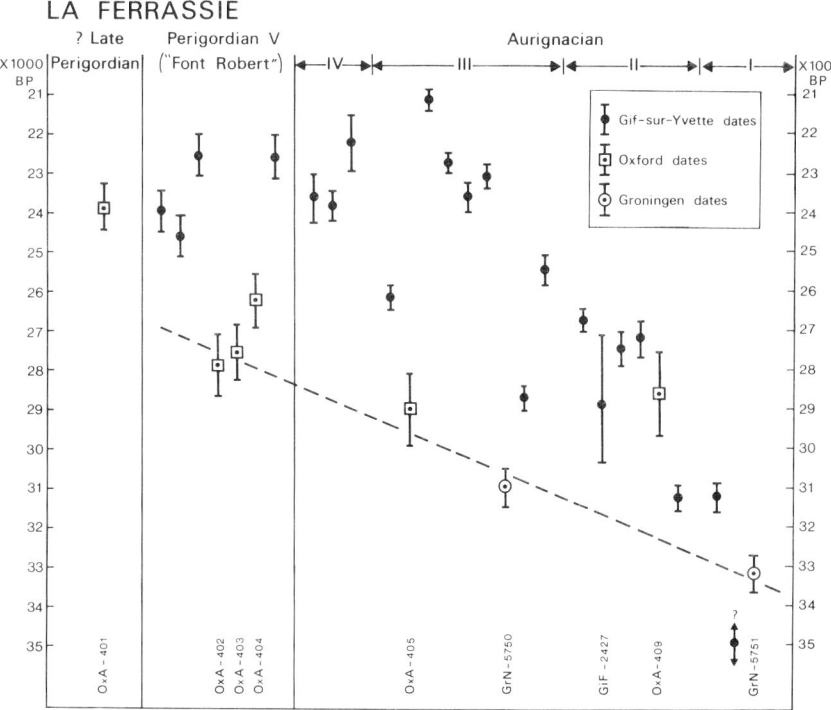

Fig. 2 Comparison of ^{14}C dates for the La Ferrassie sequence produced by the Gif-sur-Yvette, Groningen and Oxford laboratories. All of the dates are based on samples of animal bone, with the exception of Gif-2427 which is based on a sample of charcoal (see Delibrias 1984). The Gif and Groningen dates are based on the total collagen fraction of the bone samples; the Oxford dates are based on the total amino acid fraction. According to Delibrias (1984) many of the Gif dates may be affected by recent contamination of the samples; the suggested (approximate) chronology of the deposits is indicated by the broken line. For details of Oxford dates, see Table 2.

in relation to other dated samples and to the associated archaeological material. Thirdly, the selective dating of the amino acid fraction of the bone samples (again made possible by the ability to date exceptionally small samples) allows much more effective pretreatment of the samples to remove potential contamination by modern carbon. It is fair to claim that with these refinements, the radiocarbon dating of Palaeolithic sites is entering a new era. The Oxford laboratory is at present in the forefront of these developments, and may well contribute to the construction of a radically revised chronology for several aspects of the Upper Palaeolithic sequence over the course of the next few years.

ACKNOWLEDGEMENTS

We are especially indebted to H. Delporte for providing samples for dating from La Ferrassie and the Abri du Facteur, and to J-P. Rigaud for providing the samples from Le Flageolet I. For similar assistance in the submission of samples from the Abri Pataud we are grateful to Professor H.L. Movius, Professor H. de Lumley, and Mme. B. Delluc. Travel funds to assist with the collection of samples were provided by the Wenner-Gren Foundation for Anthropological Research (to H.M.B.) and by the British Academy (to P.M.).

REFERENCES

Bricker, H.M. and David, N.C., 1984, *Excavation of the Abri Pataud, Les Eyzies (Dordogne): the Perigordian VI (Level 3) assemblage*, American School of Prehistoric Research, Bulletin 34, Cambridge (Mass.): Peabody Museum, Harvard University.

David, N.C., 1985, *Excavation of the Abri Pataud, Les Eyzies (Dordogne): the Noaillian (Level 4) assemblage and the Noaillian culture in western Europe*, American School of Prehistoric Research, Bulletin 37, Cambridge (Mass.): Peabody Museum, Harvard University.

David, N.C. and Bricker, H.M., 1986, Perigordian and Noaillian in the greater Perigord, in *Regional perspectives on the Pleistocene prehistory of the Old World* (ed. O. Soffer), New York: Plenum Publishing Corporation (in press).

Delibrias, G., 1984, La datation par le carbone 14 des ossements de La Ferrassie, in *Le Grand Abri de La Ferrassie: Fouilles 1968–1973* (ed. H. Delporte), Etudes Quaternaires, Mémoire No. 7, pp. 105–108, Paris: Institut de Paléontologie Humaine.

Delporte, H., 1968, L'abri du Facteur à Tursac (Dordogne): étude générale, *Gallia Préhistoire* 11 (1), 1–112.

Delporte, H., 1984, *Le Grand Abri de La Ferrassie: Fouilles 1968–1973*, Etudes Quaternaires, Mémoire No. 7, Paris: Institut de Paléontologie Humaine.

Gillespie, R., Hedges, R.E.M. and Wand, J.O., 1984, Radiocarbon dating of bone by accelerator mass spectrometry, *J. Archaeol. Sci.* 11, 165–170.

Gowlett, J.A.J., Hedges, R.E.M., Law, I.A. and Perry, C., 1986, Radiocarbon dates from the Oxford AMS system: Archaeometry datelist 4, *Archaeometry* 28, 2, 206–221.

Laville, H., Rigaud, J-P. and Sackett, J., 1980, *Rock shelters of the Périgord*, New York: Academic Press.

Rigaud, J-P., 1982, Le Paléolithique en Périgord: les données du sud-ouest Sarladais et leurs implications, Thèse de Doctorat d'Etat ès Sciences, University of Bordeaux.

Movius, H.L. (ed.), 1975, *Excavation of the Abri Pataud, Les Eyzies (Dordogne)*, American School of Prehistoric Research, Bulletin 30, Cambridge (Mass.): Peabody Museum, Harvard University.

Movius, H.L., 1977, *Excavation of the Abri Pataud, Les Eyzies (Dordogne): Stratigraphy*, American School of Prehistoric Research, Bulletin 31, Cambridge (Mass.): Peabody Museum, Harvard University.

Vogel, J.C. and Waterbolk, H.T., 1963, Groningen radiocarbon dates IV, *Radiocarbon*, 5, 163–202.

Vogel, J.C. and Waterbolk, H.T., 1967, Groningen radiocarbon dates VII, *Radiocarbon* 9, 107–155.

Vogel, J.C. and Waterbolk, H.T., 1972, Groningen radiocarbon dates X, *Radiocarbon* 14, 6–110.

Waterbolk, H.T., 1971, Working with radiocarbon dates, *Proc. Prehist. Soc.* 37, 2, 15–33.

A.M.S. RESULTS FROM CHEDDAR GORGE – TRODDEN AND UNTRODDEN 'LIFEWAYS'

R. M. Jacobi

INTRODUCTION

Any researcher with interests in the Lateglacial archaeology of the British Isles could not fail but be envious of one slide shown at this conference. It was of part of the succession of culturally rich and geologically undistorted sediments below the rock-shelter of Laugerie Haute.

The mental comparisons triggered by equivalent exposure have undoubtedly done much to discourage most archaeologists our side of the Channel from taking very seriously the Lateglacial archaeology of Britain. This nagging impression of cultural inadequacy has in turn been linked to natural pessimism when faced with collections formed by past generations of archaeologists of whom we have frequently chosen to believe little good.

Neither reaction, however, seems useful. Britain could, perhaps, be better seen as a naturally defined sampling area for the Lateglacial archaeology of northwestern Eurasia. That its Lateglacial archaeological record is genuinely less rich or less complete than that of areas of the adjacent mainland still requires further testing, and if proven, explanation. That the full potential of collections made over a one hundred and fifty year time-span has been fully exploited is open to considerable doubt.

Indeed, optimism that structure exists amongst at least a part of the lithic material preserved within such collections is manifest in Garrod's definition of a 'Creswellian' (Garrod 1926, p. 194). Many, but not all, chipped stone technologies of Lateglacial date found in Britain appear to fit her definition comfortably. This could in turn encourage the expectation that a significant part of our Lateglacial archaeology may be chronologically discrete.

PROJECT AIMS

These are to learn more about human presence in Britain during the Lateglacial. An ability to go beyond just the chronology of this presence and begin modelling 'lifeways' for this time depends, however, upon accurate linkage of the human record to the mammal biostratigraphy of the Lateglacial. A better understanding of this biostratigraphy has, therefore, to form a second crucial component of any research design.

The perhaps unexpected corollary of this second aim is a need to know more of those faunas accumulated when local artifact discard seems *not* to have been taking place. Their composition, it could be argued, may hold clues helpful to explaining these 'gaps' in the archaeological record.

Four approaches appear useful:

(1) The dating of human fossils (*cf.* Stringer, this volume).

(2) The dating of single skeletal elements unambiguously modified during dismemberment or converted into formal artifact classes (*cf.* Cook and Barton, this volume).

 Selection for single modified skeletal elements allows realistic dating of both the local presence of that species as well as its exploitation by man. Recent papers have highlighted the difficulties of achieving either objective when conventional technology required 'bulking' of unmodified or modified organic items respectively (*cf.* Lawson 1984 and Teheux 1985).

 In cave situations, such as those which concern us here, the agencies of accumulation for faunal material can be highly varied. Radiocarbon ages can, therefore, only begin to be extrapolated with confidence to inorganic tool categories where the accompanying dated organic component is part of a series of skeletal elements whose individual surface modifications can be demonstrated as linked with a single functionally explicable pattern of butchery and processing. The possibility of mixing of a butchery sequence of one age and artifacts of another by such processes as 'mass-movement' requires to be tested for in each case.

 Finally, dates for *isolated* human or humanly modified fossils from sediments, the taphonomy of whose other organic components cannot certainly be attributed to human intervention, should be used with caution. This caution applies even where a sediment has provided evidence for human presence in the form of, for example, chipped stone tools.

 In such a context the fossil could represent food debris 'genuinely' to be associated with *human* use of that find-spot. Alternatively, it could have been scavenged by some other accumulator from any human location within its range. The two users of the find-spot sampled need not be close in time.

(3) The dating of a single animal with an embedded weapon-head or heads (*cf.* Jacobi *et al.* 1986). This should establish both the age of that animal and of its hunting. Since, however, hunting equipment from this sort of context can have been carried some distance (*cf.* Becker 1973, p. 133) such dates may refer only very uncertainly to local human presence.

(4) The dating of single fossils. Every fossil should be regarded as potentially 'stratigraphically mobile'. This mobility may be 'genuine' either within or between sediments. Effects equivalent to such mobility have also to be assumed where critical contextual information no longer survives. In either case only dating aimed at each individual species present may warn against generation of spurious community and taphonomic data — as well as untrodden 'lifeways'.

 A.M.S. technology also opens the way for dating fossils too small or too rare to be considered for conventional techniques, but which as environmental indicators may be critical to characterizing the landscapes chosen or ignored by Lateglacial hunters.

MENDIP CAVES

Mendip is regarded as a naturally defined sampling area.

 Fossil collections from Gough's Cave, Gough's Old Cave, Sun Hole, Chelm's Combe Shelter and Aveline's Hole are believed to include components capable of being 'nested' within a local mammal biostratigraphy for the Lateglacial and Earliest Holocene (Aveline's Hole). Chipped stone artifacts of Lateglacial type are recorded from Gough's Cave, Sun Hole and Aveline's Hole. There are also human fossils from all three find-spots (Stringer 1986, and this volume).

Human fossils from Aveline's Hole can be suspected as components of the only 'Mesolithic' cemetery so far identified for Britain and one of the largest in western Europe. With previous conventional determinations, however, combining human bone fragments, this claim remains to be verified.

Application of A.M.S. technology to selected items from these find-spots appeared the most helpful to the project outlined.

APPLICATIONS

Between 1927 and 1931 R.F. Parry, then agent for the Marquess of Bath at Cheddar, collected from a 'wedge' of dense sediments introduced by processes of creep and sheet wash through the mouth of Gough's Cave (Collcutt 1986) the largest sample of modified flint, chert, bone and antler objects so far recovered from a British cave locality used in Lateglacial time (Parry 1929, 1931). There were also a number of human fossils (Stringer 1986). Assuming this 'wedge' contained no hidden unconformities, its formation could have taken place within a few centuries (Collcutt 1986).

It appears valid to treat the modified flint and chert items recovered from the artificial spits into which Parry divided this sediment as sub-samples of the same lithic population (Jacobi 1986). This technology can be diagnosed as 'Creswellian' (*sensu* Garrod 1926).

Large mammal bones were also collected from this sediment 'wedge'. Both conventional (Burleigh *et al.* 1985) and accelerator dating (see below) suggest that there would be nothing unreasonable in suspecting all fossils of horse from this sediment as being chronologically inseparable. Many bones of both horse and red deer show clear cut-marks documenting patterns of both butchery and tendon removal (Parkin, *et al.* in prep.). Groups of incisions can be traced passing from one bone to another across joints. There is no evidence from the preservation condition of both organic and inorganic artifacts for their 'mass-movement' (see also comments in Collcutt 1986).

There are now the following A.M.S. results for fossils of horse with clear patterns of cut-marks (A = normal Oxford amino acid treatment; C = sample dated from BM collagen preparation) (see Gowlett *et al.* 1986, pp. 117-118):

OxA-464 Distal metapodial	12,470 ± 160 BP (A)
OxA-465 2nd phalanx	12,360 ± 170 BP (A)
OxA-589 Atlas vertebra	12,340 ± 150 BP (C)
OxA-590 Atlas vertebra (same sample as OxA-589)	12,370 ± 150 BP (A)

An unmodified fossil of horse is dated:

OxA-591 Distal metapodial	12,260 ± 160 BP (C)
OxA-592 Distal metapodial (same sample as OxA-591)	12,500 ± 160 BP (A)

A single fossil of red deer with clear pattern of cut-marks has been dated:

OxA-466 Distal metapodial	12,800 ± 170 BP (A)

This result is not significantly different from the others (Gillespie *et al.* 1985, p. 238).

The small sample sizes required for accelerator dating have meant in each case preservation of both palaeontological specimens and their surface modifications for future specialists. In turn, further sampling of these items, should it ever be necessary, remains perfectly feasible.

Extrapolation of the results for modified fossils to the chipped flint and chert collection would provide the first certain confirmation that some part of the 'Creswellian' dates to the latter part of the 'Lateglacial Interstadial' (*sensu* Lowe and Gray 1979).

Further determinations allow additions to be made to this 'Interstadial' fauna:

OxA-463 Calcaneum of Saiga antelope   12,380 ± 160 BP (A)
OxA-588 Partial phalanx *cf. Bovini*   12,030 ± 150 BP (A)
Neither fossil shows evidence for human modification.

While two 'bâtons de commandement' are clearly made of reindeer antler there is no reindeer bone from Gough's Cave with evidence for butchery or processing modification. It remains to be tested if the very few fossils of reindeer from the Cave will prove to be indistinguishable in terms of radiocarbon age from the species already dated, or to be of more recent date. The latter possibility appears to be hinted at by the conventional radiocarbon determination for a poorly contexted partial antler from Gough's Cave of:

Q-1581 = 9920 ± 130 BP (Clutton-Brock and Burleigh 1983, table 2)

Other species of whose chronology at Gough's we hope to learn more are wild boar, wild cattle (*Bos primigenius*) and arctic fox.

It is not known what proportion of Lateglacial activity at the entry to Cheddar Gorge actually took place within, rather than outside, Gough's Cave. Evidence for valley-floor activity exterior to the cave can, however, be hypothesized lost as a result of erosion during the several hundred years of increased precipitation and lowered temperatures following about 12,000 years ago. It could also be suspected, by analogy with later time, that the cave may have been 'closed' over long periods both for human use and to sediment build-up (Jacobi 1986). In either case Gough's may possess only a very partial record of Lateglacial human presence in this area.

There is no preserved Lateglacial age archaeological material from any cave in Cheddar Gorge whose typology confirms it as of a different age from that of Gough's.

Testing for human use of the 'Cheddar' area at any time different from that sampled at Gough's is clearly going to be difficult. One approach has been to seek human, or humanly modified, fossils preserved in other caves near the Gorge mouth. If not to be convincingly associated with human use of just these find-spots (see below) they may have been introduced from other residence or activity locations, some perhaps not now available for direct archaeological sampling (see above). Any of these locations could, possibly, be of an age, or ages, different from human use of Gough's Cave.

Two of three such items so far traced have been dated:

Sun Hole
OxA-535 Proximal fragment of human left ulna   12,210 ± 160 BP (A)
Gough's Old Cave
OxA-587 Modified 1st phalanx of horse   12,530 ± 150 BP (A)

There are no Lateglacial age artifacts from Gough's Old Cave and no further modified fossils. The human bone from Sun Hole recovered between 1926 and 1933–4 is not necessarily to be associated with the chipped flints from here (Tratman 1955, fig. 10) since wolf, not man, appears the major faunal accumulator (Collcutt *et al.* 1981). There is no scrap of modified bone from this cave.

Chipped flints of Lateglacial type have been found at Sun Hole (see above), Soldier's Hole (Parry 1931, pl. 12) and perhaps Flint Jack's Cave (Oakley 1958, fig. 19). In each case these cannot be distinguished typologically or technologically from those found in Gough's Cave.

CONCLUSIONS

A.M.S. results from Cheddar Gorge provide the first chronological fix anywhere in Europe for a 'Creswellian' type technology.

They allow confident identification of components of the 'Lateglacial Interstadial' fauna.

A.M.S. results appear to reinforce traditional typological approaches in suggesting human use of the entry to Cheddar Gorge during only one part of the Lateglacial.

It is hoped that further dating applied to Mendip find-spots will clarify details of mammalian biostratigraphy — including human — fo.· other parts of the Lateglacial and earliest Holocene.

Finally, it is hoped that these preliminary results will encourage investigation of mammalian biostratigraphy in other areas of Britain and with it testing for any different chronological patterns of human landscape use during the Lateglacial.

ACKNOWLEDGEMENTS

The identifications of the specimens dated are all by A.P. Currant and C.B. Stringer. The mammal fauna from Gough's Cave is at present being written up by A.P. Currant, and it is hoped to look in more detail at collections from the other Cheddar caves mentioned. We would all three like to thank C.J. Hawkes for making available the fossils from Gough's Old Cave and Sun Hole for study and dating — also for prodding us to think of new aspects of the Gorge's history. I thank my wife for spotting the Gough's Old Cave fossil.

REFERENCES

Becker, C.J., 1971, Late Palaeolithic finds from Denmark, *Proc. Prehist. Soc.* 37, 2, 131–139.

Burleigh, R., Jacobi, E.B. and Jacobi, R.M., 1985, Early human resettlement of the British Isles following the last glacial maximum: new evidence from Gough's Cave, Cheddar, *Quaternary Newsletter* 45, 1–6.

Clutton-Brock, J. and Burleigh, R., 1983, Some archaeological applications of the dating of animal bone by radiocarbon with particular reference to post-Pleistocene extinctions, in *Proc. 1st Int. Symp. on *14*C and Archaeology, Groningen, 1981, PACT* 8, 409–418.

Collcutt, S.N., 1986, Analysis of sediments in Gough's Cave, Cheddar, Somerset, and their bearing on the Palaeolithic archaeology, *Proc. Univ. Bristol Spelaeol. Soc.* 17, 2 (for 1985), 129–140.

Collcutt, S.N., Currant, A.P. and Hawkes, C.J., 1981, A further report on the excavations at Sun Hole, Cheddar, *Proc. Univ. Bristol Spelaeol. Soc.* 16, 1, 21–38.

Cook, J. and Barton, R.N.E., 1986, Dating Late Devensian — Early Flandrian barbed points, this volume.

Garrod, D.A.E., 1926, *The Upper Palaeolithic Age in Britain*, Oxford: Clarendon Press.

Gillespie, R., Gowlett, J.A.J., Hall, E.T., Hedges, R.E.M. and Perry C., 1985, Radiocarbon dates from the Oxford AMS system: Archaeometry datelist 2, *Archaeometry* 27, 2, 237–246.

Gowlett, J.A.J., Hall, E.T., Hedges, R.E.M., Perry, C., 1986, Radiocarbon dates from the Oxford AMS system: Archaeometry datelist 3, *Archaeometry* 28, 1, 116–125.

Jacobi, R.M., 1986, The Lateglacial archaeology of Gough's Cave at Cheddar, in *Recent studies in the Palaeolithic of Britain and its nearest neighbours* (ed. S.N. Collcutt), Sheffield: J.R. Collis Publications, Department of Archaeology and Prehistory, Sheffield University.

Jacobi, R.M., Gowlett, J.A.J., Hedges, R.E.M. and Gillespie, R., 1986, Accelerator mass spectrometry dating of Upper Palaeolithic finds with the Poulton elk as an example, in *Studies in the Upper Palaeolithic of Britain and northwest Europe* (ed. D.A. Roe), Oxford: B.A.R. International Series (in press).

Lawson, T.J., 1984, Reindeer in the Scottish Quaternary, *Quaternary Newsletter* 42, 1–5.

Lowe, J.J. and Gray, J.M., 1979, The stratigraphic subdivision of the late glacial of N.W. Europe: a discussion, in *Studies in the Lateglacial of north-west Europe* (eds. J.J. Lowe, J.M. Gray and J.E. Robinson), pp. 157–175, Oxford: Pergamon Press.

Oakley, K.P., 1958, The antiquity of the skulls reputed to be from Flint Jack's Cave, Cheddar, Somerset, *Proc. Univ. Bristol Spelaeol. Soc.* 8, 2 (for 1957–8), 77–82.

Parkin, R.A., Rowley-Conwy, P. and Serjeantson, D., in prep., Palaeolithic exploitation of horse and red deer at Gough's Cave, Somerset.

Parry, R.F., 1929, Excavation at the Caves, Cheddar, *Proc. Somerset Archaeol. and Nat. Hist. Soc.* 74, 2 (for 1928), 102–121.

Parry, R.F., 1931, Excavations at Cheddar, *Proc. Somerset Archaeol. and Nat. Hist. Soc.* 76, 2 (for 1930), 46–62.

Stringer, C.B., 1986, The hominid remains from Gough's Cave, *Proc. Univ. Bristol Spelaeol. Soc.* 17, 2 (for 1985), 145–152.

Stringer, C.B., 1986, Direct dates for the fossil hominid record, this volume.

Teheux, E., 1985, Nouvelles fouilles sur le site magdalénien de Chaleux, *Cahiers de Préhistoire Liègeoise* I, 95–103.

Tratman, E.K., 1955, Second report on the excavations at Sun Hole, Cheddar: the Pleistocene levels, *Proc. Univ. Bristol Spelaeol. Soc* 7, 2 (for 1954–5), 61–70.

DATING LATE DEVENSIAN – EARLY FLANDRIAN BARBED POINTS

J. Cook and R. N. E. Barton

INTRODUCTION

Barbed bone and antler points have long been recognised as the most characteristic artefacts of the early Post-glacial period in Britain. However, although they have come to be regarded as 'marker' or 'type' fossils, they are poorly dated because they often occur as single finds, lacking well-documented contexts and sometimes without definite provenances. As a result, relative dating schemes stemming from typological studies have produced conflicting chronologies and interpretations (Clark and Godwin 1956; Wymer *et al*. 1975) and although some of these points could be said to indicate the reappearance of people in Britain following the last glacial maximum of the Devensian, the timespan of their production has remained ambiguous. To alleviate this situation, the present authors are collaborating on a new study of these points which involves direct dating and a review of their morphology and technology, including experimental work and microscopy. Direct dating of the points themselves could only be done using the sampling and preparation techniques and better dating resolution developed by the Oxford Radiocarbon Accelerator Unit because conventional radiocarbon techniques would have involved the partial or, in some cases, total destruction of the specimens. The first results of this dating programme are presented here.

MATERIAL DATED

The first group of material submitted for dating includes the barbed bone and antler points from Sproughton, Suffolk, and antler points from Earls Barton, Northamptonshire and Waltham Abbey, Essex. These pieces were selected because they all have some contextual information and show ancient break surfaces which could be drilled for samples with a minimum of risk. About 0.3 g of material have been taken from each piece and the holes unobtrusively plugged with an inert filler by Dr John Gowlett of the Accelerator Unit. Dates are now available for the first three specimens noted above.

RESULTS

The Sproughton points

The two uniserial barbed points from Devils Wood Pit in Sproughton near Ipswich (NGR: TM134443) were found in sand and gravel deposits of a former channel of the River Gipping during commercial working of the sediments (Wymer *et al*. 1975). The bone point was recovered from gravel about 15 feet from the surface and on the basis of conventional radiocarbon dates obtained on organic material from the sequence was expected to date between about 11,300 and 9800 BP or within pollen Zone III (Wymer *et al.* 1975). The antler point came from a fine gravel in cross-bedded sediments about 0.8 m from the top of the sequence. On the basis of its long tang, it was suggested (Wymer *et al.* 1975) that this point was of Post-glacial, probably Zone IV, age. However, the accelerator dates given below indicate that the points are remarkably similar in age:

OxA-517 Sproughton bone point:	$10,910 \pm 150$ BP
OxA-518 Sproughton antler point:	$10,700 \pm 160$ BP

These dates are comparable with those obtained for late Zone II and early Zone III deposits at Pitstone, Buckinghamshire (Evans, this volume) and both points may now be regarded as two of the few well-substantiated archaeological finds from the late glacial period.

Earls Barton point

This unfinished uniserial barbed antler point (Fig. 1) was recovered from gravel deposits of the lower floodplain terrace of the River Nene, east of Northampton between Clifford Hill and Doddington (NGR: 875625) during commercial extraction of the sediments. This site, now flooded, consisted of an organic silt stratified between gravels. The barbed point probably came from the lower gravel but this is uncertain. Other finds from this deposit include remains of woolly rhinoceros, *Coelodonta hemitoechus*, dated by conventional radiocarbon to $25,500 \pm 650$ BP (BM 2074C: Burleigh *et al.* 1984, p. 61; see also Gillespie *et al.* 1984) and a 'Lyngby-axe' made on a reindeer antler which has been submitted for accelerator dating at Oxford. The presence of these faunal species suggested a possible late glacial date for the point although on stylistic grounds it was estimated to be of early postglacial, Zone IV age. The accelerator date of 9240 ± 160 BP (OxA-500) has confirmed the Zone IV dating of the piece.

CONCLUSIONS

These preliminary results indicate that small fragile bone and antler objects can be successfully dated using minimally destructive sampling techniques developed in Oxford. This will make it possible to include undamaged specimens from sites such as Thatcham, Berkshire and Brandesburton, Yorkshire, on the list of specimens now being sought for dating. The first dates achieved for the barbed points show that it should be possible to ascertain when people occupied certain areas of Britain by directly dating their bone and antler work. Combined with new work on the morphology and technology of the points, further dates should provide a more secure relative chronology of barbed point types. These dates will also allow more reliable comparisons of cultural material and environmental data and contribute to our understanding of the Upper Palaeolithic to Mesolithic transition.

ACKNOWLEDGEMENTS

Thanks are due to Mr Graham Teal and his son Adrian for allowing the barbed point and Lyngby-axe from Earls Barton to be included in this project and Mrs Hilary Feldman of Ipswich Museum for permission to sample the Sproughton points.

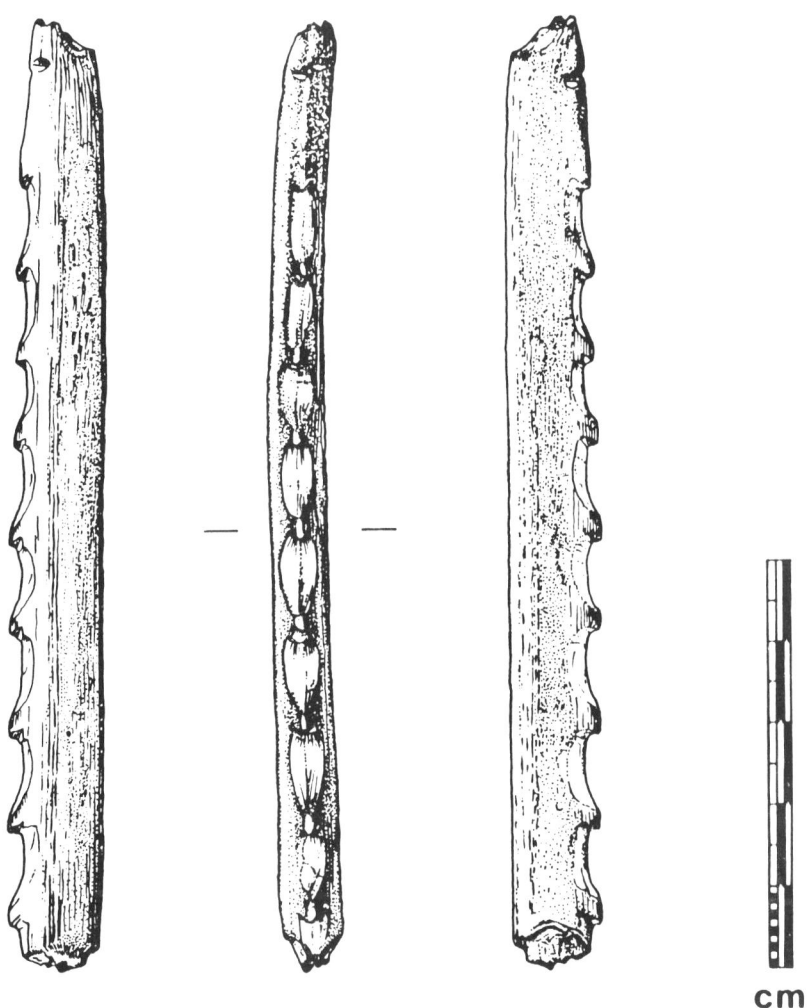

Fig. 1   Barbed antler point from Earls Barton, Northamptonshire.

REFERENCES

Burleigh, R., Ambers, J. and Matthews, K., 1984, British Museum Natural Radiocarbon Measurements XVII, *Radiocarbon* 26, 1, 59–74.

Clark, J.G.D. and Godwin, H., 1956, A Maglemosian site at Brandesburton, Holderness, Yorkshire, *Proc. Prehist. Soc.* 22, 6–22.

Gillespie, R., Hedges, R.E.M. and Wand, J.O., 1984, Radiocarbon dating of bone by accelerator mass spectrometry, *J. Archaeol. Sci.* 11, 165–170.

Wymer, J.J., Jacobi, R.M. and Rose, J., 1975, Late Devensian and Early Flandrian barbed points from Sproughton, Suffolk, *Proc. Prehist. Soc.* 41, 235–241.

RADIOCARBON DATES FROM THE PITSTONE SOIL
AT PITSTONE, BUCKINGHAMSHIRE

J. G. Evans

At Pitstone, Buckinghamshire, in two scarp-slope coombes of the Chalk, a palaeosol is stratified between deposits of chalky colluvium (Evans 1966). The local geomorphology, biostratigraphy (land Mollusca) and lithostratigraphy suggest an age within the Windermere Interstadial of the Devensian Lateglacial, and the palaeosol has been named the Pitstone Soil (Rose *et al.* 1985). Measured drawings of the exposed section are reproduced here (Fig. 1).

In the surface 1.5 cms of the soil small charcoal fragments are present and many of the shells are calcined, indicating intense local burning. The fragmentary and generally dispersed nature of the charcoal makes this site ideal for accelerator radiocarbon dating. Accordingly, samples were collected by members of the laboratory and the author on 10th July 1984 from the exposed North-east Section (Evans 1966; fig. 1), and dated by the Oxford University Radiocarbon Accelerator Unit.

The dates are as follows:

OxA-415	Pitstone 4 charcoal	10,900 ± 130 BP
OxA-426	Pitstone 3 charcoal	10,410 ± 150 BP
OxA-427	Humic acid extracted from Pitstone 3	10,400 ± 150 BP
OxA-428	Pitstone 2 charcoal	8800 ± 140 BP
OxA-429	Humic acid extracted from Pitstone 2	9400 ± 140 BP

The laboratory commented as follows: "OxA-415 provides the most reliable date. OxA-426 was much less well-preserved charcoal, but the agreement of the humics, OxA-427, suggests that it is a good date. OxA-428 is very significantly younger, and the older humic acid date, OxA-429, suggests that this is an intrusive sample. We have a number of examples from other sites where younger intrusive samples 'gain' an older humic acid date from the surrounding older sediment."

These comments are in agreement with the expected results. Thus, OxA-415, 10,900 ± 130BP, falls close to the end of the Windermere Interstadial (*c.* 10,750 BP) and confirms the ascription of the Pitstone Soil to this period. The dates of 10,410 ± 150 BP and 10,400 ± 150 BP (OxA-426 and 427) are slightly less satisfactory, falling within the Loch Lomond Stadial. There is, however, overlap at two standard deviations with OxA-415, and they certainly belong to the same general period.

OxA-428 and OxA-429, on the other hand, are far too young to belong to the Devensian Lateglacial, and must be rejected.

Two other radiocarbon dates from similar contexts and in association with similar land molluscan assemblages may be noted. They are:

91

Q-463   Charcoal from a palaeosol at Dover Hill, Folkestone, Kent   11,934±210 BP
         (Kerney 1963)
Q-618   Organic detritus mud, Borehold III, Brook, Kent   11,900±160 BP
         (Kerney *et al.* 1964)

Both dates are from contexts ascribed to the Windermere Interstadial on molluscan and stratigraphical grounds. The radiocarbon dates indicate a time at the very beginning of that period. They are significantly earlier than the Pitstone dates, with no overlap at two standard deviations.

A much earlier date of 13,180±230 (Q-473) comes from charcoal in a weakly developed palaeosol within chalky colluvium at Holborough, Kent (Kerney 1963), ascribed to the Early Windermere Interstadial (Zone Ib) of the Devensian Lateglacial.

At Pitstone, in addition to the main charcoal horizon at the soil surface, there is an intermittent humic horizon with charcoal a few centimetres below (fig. 1). There were thus several burning episodes within the Devensian Lateglacial in the chalklands of southern England. Whether these were of human origin as has occasionally been suggested or the result of natural fires is unknown, but it should be noted that none of the dated horizons has associated archaeology. Nevertheless, Upper Palaeolithic communities were present at this time in southern England. For example, the two dates for the Sproughton barbed points reported on in this volume (Cook and Barton) of 10,910±150 BP (OxA-517) and 10,700±160 BP (OxA-518) are indistinguishable from the best of the Pitstone dates (OxA-415, 426 and 427).

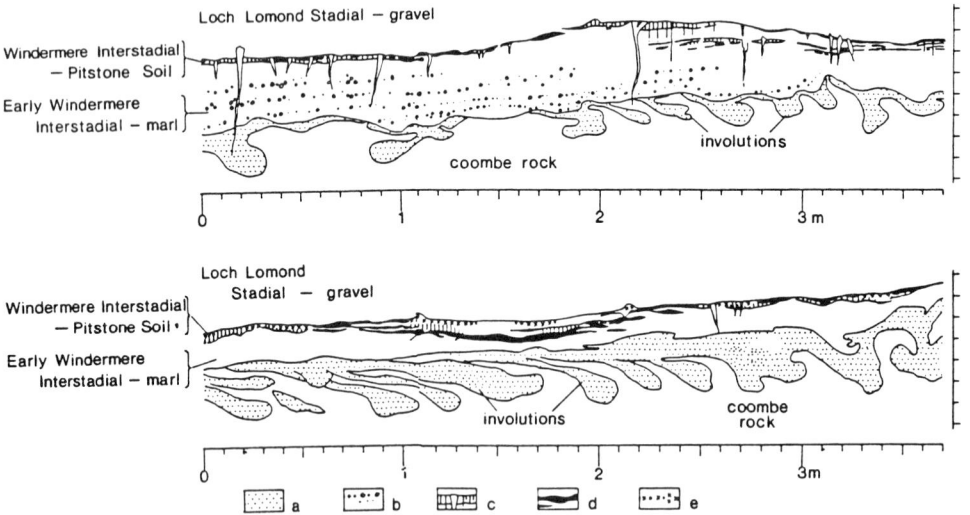

Fig. 1   *Representative sections of the Pitstone Soil: a=chalky clay; b=chalk mud and fine rubble (marl); c=humic chalk mud with cracks; d=black humus with charcoal; e=pale brown, loose friable loam with iron staining. Vertical and horizontal scale in 0.1 m intervals. (The two sections are 8.8 m apart.)*

ACKNOWLEDGEMENTS

I would like to thank the Oxford University Radiocarbon Accelerator Unit for carrying out the measurements. Tunnel Cement Ltd., Pitstone Works, kindly gave access to the site.

REFERENCES

Cook, J. and Barton, R. N. E., 1986, Dating Late Devensian — Early Flandrian barbed points, this volume.

Evans, J. G., 1966, Late-Glacial and Post-Glacial subaerial deposits at Pitstone, Buckinghamshire, *Proc. Geologists' Assoc.* 77, 3, 347–364.

Kerney, M. P., 1963, Late-Glacial deposits on the Chalk of South-East England, *Phil. Trans. R. Soc.* (B) 246, 203–254.

Kerney, M. P., Brown, E. H. and Chandler, T. J., 1964, The Late-Glacial and Post- Glacial history of the Chalk escarpment near Brook, Kent, *Phil. Trans. R. Soc.* (B) 248, 135–204.

Rose, J., Boardman, J., Kemp, R. A. and Whiteman, C. A., 1985, Palaeosols and the interpretation of the British Quaternary stratigraphy, in *Geomorphology and soils* (eds. K. S. Richards, R. R. Arnett and S. Ellis), London: George Allen and Unwin, pp. 348–375.

COMPLEMENTARITY OF CONVENTIONAL AND ACCELERATOR DATING: EXAMPLES IN PLEISTOCENE EXTINCTIONS

Richard Burleigh

The topic of this paper is late Pleistocene and early Holocene mammalian extinctions with particular reference to the British Isles, but with some mention also of Mediterranean islands, and of an opposite aspect of the extinction process, that is introductions.

For some years the British Museum radiocarbon laboratory, in collaboration with the British Museum (Natural History), has had a programme for dating the extinction of larger terrestrial mammals in Britain following the end of the last glaciation. An essential condition of this project has been the dating of only those bones that could be positively identified to species, and each measurement has been based whenever possible on a single element of the skeleton, usually one of the long-bones, of a single individual (Clutton-Brock and Burleigh 1983). Examples of species that we have been particularly interested in are the reindeer, the aurochs, and the wild horse. Dates are based on the preserved 'collagen', separated by demineralization of bone with dilute acid. This has been converted to benzene and its age determined by the conventional and well-established liquid scintillation technique. For these measurements, in order to achieve a reasonable error of ± 100 years or so, some 5 grams of benzene are needed. This requires about 100 grams of bone to begin with, given reasonably well-preserved material, and the requirements for antler are similar. The aim is to try to establish the latest date of survival of each extinct species investigated and, conversely, the earliest appearance of introduced taxa. The fragmentary character of the fossil record and the inevitable incompleteness of the archaeological record combine to make this difficult to achieve in practice, but we can hope to obtain a pattern of dates approximating to the true picture for each of these processes. That there are still relatively few dates results from the intermittent availability of suitable specimens.

Extinctions often have multiple causes, but in Britain at least we can probably assume that as far as the larger mammals are concerned, human interference has been a major, if not the dominant, factor in the prehistoric period. As soon as the sea-level rose in the early Holocene, separating the British Isles from continental Europe, populations of the larger species, some of them, such as the elk, *Alces alces*, probably never very abundant, could no longer be replenished and their ultimate fate was sealed. Some, perhaps more adaptable species did, it is true, survive until quite late. The latest radiocarbon date for brown bear, for example, is around 700 bc from the north of Scotland (BM-724, 2673 ± 54 BP: Burleigh *et al.* 1976, p. 30), and other species such as the wolf and the wild boar survived into historic times. Another large mammal, the red deer, that has traditionally enjoyed a measure of protection, has been present in Britain more or less continuously since the Pleistocene.

The wild horse, *Equus ferus*, forms a particularly interesting part of our investigation and here the accelerator has already made an important contribution. The series of dates listed in Table 1 for remains of wild horse from Gough's Cave, Cheddar, was actually dated for

Table 1

Comparison of radiocarbon dates for remains of wild horse, *Equus ferus*, from Gough's Cave, Cheddar, obtained by the conventional and accelerator techniques, respectively (5570 year half-life).

Spit (1)	(BM-2183-2188)	collagen (2)	amino acids (3)
10	12,120 ± 120	OxA-589 12,340 ± 150	OxA-590 12,370 ± 150
12	12,020 ± 120		
13	11,970 ± 230		
14	12,240 ± 220		
16	12,070 ± 170	OxA-591 12,260 ± 160	OxA-592 12,500 ± 160
18	12,160 ± 210		

(1) the remains dated were: an atlas vertebra from spit 10, a calcaneum from spit 12, and metapodia from spits 13–18 (see Burleigh *et al.* 1985; Burleigh *et al.*, in press, b).
(2) collagen separated from these bones at the BM laboratory.
(3) amino acids extracted at the Oxford laboratory from collagen prepared at the BM laboratory (2).
The accelerator dates for collagen are on average 200 ± 150 years older than the equivalent conventional dates and those for the separated amino acids are 330 ± 160 years older. The errors of the BM-dates are based on counting statistics (+ 10), whilst those of the OxA-dates are dominated by the uncertainty in estimating background.

archaeological reasons (Burleigh *et al.* 1985; Burleigh *et al.*, in press, b), but has also provided a closely coherent series for the extinctions programme. Some of these remains have also been dated by the accelerator and, as shown in the same table, have given results that are on average a little older. Other related samples which had cut-marks resulting from butchery and were too important to destroy totally, are also older by some 400 years (Table 2), perhaps as a result of the more searching preliminary chemistry that can be used for the preparation of the very much smaller samples needed for dating by the accelerator technique. In Table 3 the dates so far obtained by the conventional and accelerator techniques for the latest survival of the reindeer, *Rangifer tarandus*, in Britain are given.

An instance of an extinction on a Mediterranean island is that of the enigmatic ruminant *Myotragus balearicus*, on Mallorca. This animal was long considered to have become extinct during the Pleistocene, but our dating shows that it survived into the Neolithic period when it was presumably finally eliminated as a competitor for food with introduced sheep and goats (Burleigh and Clutton-Brock 1980). The principal reason for mentioning it here is that the date is based on the very diagnostic bones of *Myotragus*, emphasizing the point made earlier about the need for very precise investigation of the bones used in this type of investigation. A difficulty encountered with outwardly well-preserved bone from hotter climatic regions is, however, that of poor preservation of collagen and here the accelerator has a major role to play. Such bone when demineralized with dilute acid often dissolves away completely. The resulting acid solute does contain some of the original amino acids deriving from the thermal breakdown of collagen *in situ*, but these cannot generally be recovered in sufficient quantity for conventional dating. In another investigation we have carried out into the dating of early domestication of equids (the ass and the horse) in the Near East the accelerator has already provided us with successful dates for such material, which could not otherwise have been directly dated (Burleigh *et al.*, in press, a).

Table 2

Radiocarbon dates for remains from Gough's Cave, Cheddar, obtained by the accelerator technique (5570 year half-life).

Spit (1)	Material (2)	Lab. No.	Date (BP)
113	amino acids from metatarsal of red deer	OxA-466	12,800 ± 170
14	amino acids from calcaneum of Saiga antelope	OxA-463	12,380 ± 160
14	amino acids from 2nd phalanx of horse	OxA-465	12,360 ± 170
18	amino acids from metacarpal of horse	OxA-464	12,470 ± 160

(1) see Burleigh, Jacobi and Jacobi, 1985; Burleigh *et al.*, in press, b).
(2) the remains of red deer and horse showed cut-marks; the Saiga antelope calcaneum was too small to be dated other than by the accelerator technique.

These remains were from the same context as those listed in Table 1 dated by the conventional technique. Statistical analysis (by Dr M.N. Leese, Research Laboratory, The British Museum) shows that the weighted mean of the dates listed in Table 2 (12,500 ± 80 BP) is some 400 years older than that of the equivalent conventional dates (12,090 ± 70) listed in Table 1.

Table 3

Selected radiocarbon dates (1) for the late survival of reindeer *Rangifer tarandus*, in Britain, obtained by the conventional and accelerator techniques (5570 year half-life).

Site	Material	Lab. No.	Date (BP)
Darent gravels	antler (collagen)	BM-1674	9760 ± 70
Gough's Cave,	antler (collagen)	Q-1581	9910 ± 130 Layer 8
Soldier's Hole,	metapodial (collagen)	BM-2249	9930 ± 210 Layer 3
Chelm's Coombe,	metapodial (collagen)	BM-2318	10,190 ± 130 Area A, Spit 12
Ossom's Cave	mandible (amino acid from collagen)	OxA-631	10,780 ± 160 Layer C
Ossom's Cave	mandible (amino acid from collagen)	OxA-632	10,600 ± 140 Layer C

(1) for other, comparable dates, see Clutton-Brock & Burleigh (1983).

The very much reduced scale of destructiveness in obtaining an accelerator date is also very advantageous. A fragmentary horse mandible now in the British Museum collection, from Kendrick's Cave on the Great Orme's Head, Llandudno, North Wales, marked with a chevron pattern and one of the very few examples of prehistoric mobiliary art found in Britain (Sieveking 1971), has been dated by the accelerator to 10,000 ± 200 BP (OxA-111), in excellent agreement with our other late dates for horse obtained by the conventional technique (Clutton-Brock and Burleigh, in press).

In conclusion some general points can be made. Firstly, though the dating of extinctions is clearly the kind of investigation to which conventional techniques of radiocarbon dating are well suited, the accelerator technique also has a very definite role to play. This is not a problem in which very high precision of dating is sought; in other words normal precision only is needed and the accelerator is clearly well able to produce this. Secondly, the advantage of being able to date much smaller samples is very great. Many specimens of interest to this investigation show evidence of human modification and are far too valuable to destroy in totality and yet it is crucial to have direct dates. Here again the accelerator can make a major contribution. Thirdly and lastly, this project has suffered up till now from an inability to obtain direct dates on not only small or unique specimens, but also smaller animal species such as rodents and insectivores whose distribution is often the direct result of human intervention. The accelerator will allow us to overcome this and perhaps, in due course, to enlarge the environmental picture further still by the inclusion of selected invertebrate remains such as insects and molluscs.

REFERENCES

Burleigh, R. and Clutton-Brock, J., 1980, The survival of *Myotragus balearicus* Bate, 1909, into the Neolithic on Mallorca, *J. Archaeol. Sci.* 7, 385–388.

Burleigh, R., Clutton-Brock, J. and Gowlett, J.A.J., in press, a, Early domestic equids in Egypt and Western Asia: an additional note, in *Equids in the Ancient World*, Vol. 2 (eds. R. Meadow and H-P. Uerpmann), Tubingen: TAVO.

Burleigh, R., Currant, A., Jacobi, E. and Jacobi, R., in press, b, A note on some British late Pleistocene remains of horse (*Equus ferus*), in *Equids in the Ancient World*, Vol. 2 (eds. R. Meadow and H-P. Uerpmann), Tubingen: TAVO.

Burleigh, R., Hewson, A. and Meeks, N., 1976, British Museum natural radiocarbon measurements VIII, *Radiocarbon* 18, 1, 16–42.

Burleigh, R., Jacobi, E.B. and Jacobi, R.M., 1985, Early human resettlement of the British Isles following the last glacial maximum: new evidence from Gough's Cave, Cheddar, *Quaternary Newsletter* 45, 1–6.

Clutton-Brock, J. and Burleigh, R., 1983, Some archaeological applications of the dating of animal bone by radiocarbon with particular reference to post-Pleistocene extinctions, in *Proc. 1st Int. Symp. on C-14 and Archaeology, Groningen, Netherlands, 1981* (eds. W.G. Mook and H.T. Waterbolk), pp. 409–419, *PACT* 8.

Clutton-Brock, J. and Burleigh, R., in press, The mandible of a Mesolithic horse from Seamer Carr, Yorkshire, England, in *Equids in the Ancient World*, Vol. 2 (eds. R. Meadow and H-P. Uerpmann), Tubingen: TAVO.

Sieveking, G. de G., 1971, The Kendrick's Cave mandible, *British Museum Quarterly* 35, 230–250, pls. LXXXIV–LXXXVI.

DATING RESULTS FROM PALAEOLITHIC SITES AND PALAEOENVIRONMENTS IN EPIRUS (NORTH-WEST GREECE)

G. N. Bailey, C. S. Gamble, H. P. Higgs, C. Roubet, D. P. Webley, J. A. J. Gowlett, D. A. Sturdy and C. Turner

ARCHAEOLOGICAL BACKGROUND

The Epirus region of North-west Greece is rich in Palaeolithic finds, most of which were discovered in a series of surveys and excavations in the 1960s (Dakaris *et al.* 1964, Higgs and Vita-Finzi 1966, Higgs *et al.* 1967, Sordinas 1969). They include the limestone rockshelters of Asprochaliko in the lower Louros Valley, with a long sequence of Mousterian and Upper Palaeolithic industries, Kastritsa on the shores of Lake Ioannina (with late Upper Palaeolithic only), and Grava on the island of Corfu (also late Upper Palaeolithic). There are also numerous open-air sites consisting of flint artefacts eroding out of old land surfaces composed of red sediments. The flints are often of Mousterian type and the largest such concentrations are in the coastal lowlands of Epirus and Corfu, the most famous being the site of Kokkinopilos, near Asprochaliko, where Eric Higgs discovered the first evidence of Palaeolithic occupation in 1962. One of the principal outcomes of the Higgs investigations, apart from the establishment of the main outlines of a Palaeolithic sequence for the region, was the proposal that the coastal and inland sites should be interpreted as seasonal camps forming complementary elements in a regional pattern of settlement and land use.

The Klithi project was originally inspired by the need to provide a full analysis and publication of the Palaeolithic excavations carried out in the 1960s at the sites of Kokkinopilos, Asprochaliko and Kastritsa, and by the wish to reexamine the interrelations between sites, and between sites and palaeoenvironments, implied by the hypothesis of seasonal mobility with which the 1960s work culminated (Higgs *et al.* 1967). Between 1979 and 1982 we re-examined the large store of finds in the Ioannina Museum, revisited site locations in the field, and collected fresh dating samples from the Asprochaliko section (Bailey *et al.* 1983a, 1983b). Since 1983 work has been focussed on excavation at the Klithi rockshelter (discovered in 1979 (Bailey *et al.* 1984, in press)) and on palaeoenvironmental work in the vicinity of Klithi.

FIELDWORK STRATEGY

The current strategy is based on the premise that pre-agricultural settlement systems are regional in scale, and that to achieve a correct understanding of Palaeolithic social and economic change from analysis of excavated data it is necessary to control for three variables:

(1) intra-site variation — the point here being that contemporaneous material cannot be assumed to be distributed at random, even within the confines of a rockshelter, but is more

Table 1

LIST OF RADIOCARBON DATES FROM KLITHI

Klithi Rockshelter

OxA-136	Amino acid from bone collagen	16,300 ± 400 BP
	Rectangle W32, Spit 77, base of section	
OxA-137	Amino acid from bone collagen	17,000 ± 400 BP
	Rectangle W32, Spit 77, base of section	
OxA-501	Charcoal under stone	260 ± 100 BP
	Rectangle P27, subsurface	
OxA-502	Charcoal fragments from hearth	12,300 ± 200 BP
	Rectangle Q22, Spit 16, Layer 15, H1	
OxA-542	Burnt ibex bone from hearth	10,420 ± 150 BP
	Rectangle S20, Spit 16, Layer 14, H3	
OxA-747	Charcoal fragments	3560 ± 1000
	Rectangle S21, Spit 16, Layer 14, B1203	
OxA-748	Charcoal fragments	101.8 ± 1% modern
	Rectangle P26, Spit 11–12, Layer 14, B6002	
OxA-749	Charcoal fragments	14,200 ± 200
	Rectangle Q24, Spit 15, Layer 16, B4809	
OxA-750	Charcoal fragments	14,060 ± 200
	Rectangle P24, Spit 16, Layer 16, B5615	

Klithi Environment

OxA-191	Charcoal from sediment exposed in river cutting of Voidomatis, 100 m downstream from Kalpaki-Konitsa road bridge	1000 ± 150 BP
OxA-192	Charcoal, same provenance as OxA-191	800 ± 100 BP
OxA-351	Charcoal from hearth in old river sediments *c.* 8 m above present river level, 4 km downstream from Klithi	11,000 ± 200 BP
OxA-352	Humic acids from charcoal of OxA-351	15,840 ± 250 BP
OxA-353	Charcoal from same hearth as OxA-351	10,700 ± 200 BP
OxA-372	Humic acids from charcoal of OxA-353	14,600 ± 340 BP
OxA-512	Lake Gramousti, plant macrofossil from core at 11.58 m–11.595 m depth	7270 ± 120 BP
OxA-513	Rezina marsh, plant macrofossil from core at 8.62 m–8.635 m depth	5000 ± 100 BP

likely to be segregated into localised areas of differential activity, discard or preservation; (2) inter-site variation — the point here being that individual sites are unlikely to provide a representative picture of regional patterning, but are more likely to represent particular functions or activities (seasonal or otherwise), and even collectively to represent only the preserved fragment of the original pattern; (3) palaeoenvironmental context — both at the local scale (in relation to individual sites), and at the regional scale (in relation to the total site distribution).

Because of permit restrictions, archaeological work is at present confined to excavation at Klithi, with palaeoenvironmental work ranging more widely, so that results bear mainly on points (1) and (3) above, but also indirectly on point (2) by providing an additional site-based archaeological data set for comparison with the material from Asprochaliko and Kastritsa. AMS dating is of central importance in relation to all three points: (a) because of the residual nature of preserved organic material, especially in non-archaeological sediments; (b) because critical parts of the Palaeolithic sequence are at the lower limit of effective radiocarbon dating by conventional techniques; (c) above all because of the need for multiple correlations, and hence multiple dates, between deposits and features within-site, between archaeological sites, and between archaeological sites and off-site sediments. Dates are listed in Table 1.

KLITHI: WITHIN-SITE CORRELATIONS

Klithi is the largest known rockshelter in the region and offers ample opportunity for examination of lateral variation (Fig. 1). The deposits have not yet been excavated to bedrock, but have been exposed so far to a maximum depth of 3 m below surface. They are stratified screes formed by cold climate conditions, contain a microlithic flint industry of late glacial type, and faunal remains of ibex and chamois consistent with the local environment under a late glacial climate. In the lower part of the deposits the increased size of limestone fragments suggests a more severe climate. An expected maximum glacial date (20,000 BP to 16,000 BP) is confirmed by the two independent readings from stratigraphically identical provenances at the base of the section (OxA-136 and OxA-137) (Gillespie *et al.* 1985).

The top of the scree deposit appears to have been truncated and is overlain unconformably by recent deposits of goat dung. There is no indication from the screes themselves (such as surface weathering), from the flint industry (such as geometric

Fig. 1   Excavation grid at Klithi

microliths) or from the fauna (such as temperate species) to indicate evidence of Holocene occupation. The predicted date for the top of the scree deposit is therefore > 10,000 BP, and this is confirmed by the two available final Pleistocene readings (OxA-502 and OxA-542).

These two dates are also relevant to the interpretation of within-site spatial variation since they come from the back area of the rockshelter, where the distribution of materials and deposits is strongly influenced by cultural activities in relation to a major hearth area (Fig. 2). At the centre of this area are thick ashy deposits with very few finds (Layer 14 H3 and Layer 14 H4), and around its edge are thinner or less ashy deposits with more finds (Layer 13, Layer 14 L1, Layer 15 H1). These are surrounded by scree deposits rich in flint artefacts and bone fragments (Layer 15), with an especially dense band of debris around the front edge with many large pieces of stone, flint and bone. Beyond that again is a less stony scree deposit with very few finds (Layer 17).

The relative age of the dated samples is consistent with their stratigraphic position: the older (OxA-502) is from Layer 15 H1, a wedge of ashy deposit stratified between Layer 15 screes. The younger (OxA-542) is from the Layer 14 H3 ashy deposit, whose upper surface was in direct contact with the overlying goat dung. The time span of *ca.* 2000 years between the two dates is surprisingly large, given that they refer to material separated horizontally by only 3 m and vertically by 5 cm maximum. This argues in favour of two hypotheses: (1) that the centre of the hearth was cleaned out from time to time to make way for fresh fuel, the older material being spread out to the peripheries; (2) that this area of the site was repeatedly used in a similar way over long periods.

The spatial distribution of the various deposits and features also argues for differential use of particular locations, or differential discard of materials, around the hearth area. However, there is a relatively slow average rate of scree deposition (*ca.* 1 mm per year), for this site, and the distribution mapped in Figure 2 also represents a horizontal plane through deposits with a slight dip from south to north and west to east. Both these points are confirmed by the recently determined dates (OxA-749 and OxA-750) and urge caution in interpretation. What appears to be a penecontemporaneous spatial distribution on a horizontal surface might in fact be a stratigraphic sequence of successive deposits with a time-depth of as much as 4,000 years according to the radiocarbon dates. Thus the pattern visible on a horizontal surface might indicate differential use of the site as a whole at successive periods, rather than evidence for differential use of localised areas within the site during a single period of occupation.

The hiatus of deposition between the late glacial screes and recent goat dung, of the order of 8000 to 10,000 years, can be explained in two ways: (1) there was a general hiatus in the use of the site; (2) deposits later than 10,000 BP were removed by recent shepherds to level the surface. Hypothesis (1) is quite plausible, since the site is in a remote and inaccessible gorge, made more difficult of access during the Holocene by thicker vegetation cover and lowering of the local river channel (see below). The villages on the heights above Klithi (notably Klithonia), from which shepherds used to descend to Klithi with their animals for winter shelter, are refuge villages established during the Ottoman occupation, according to local historical records in the 17th century AD, but at any rate not earlier than the 16th century AD. If hypothesis (2) were correct, we would expect at least some traces of post-Palaeolithic activity, for example small pieces of charcoal from later prehistoric fires, to have remained mixed in with material on the surface of the scree deposits. OxA-747 provides a tantalising hint of such evidence. OxA-501, on the other hand, is a small

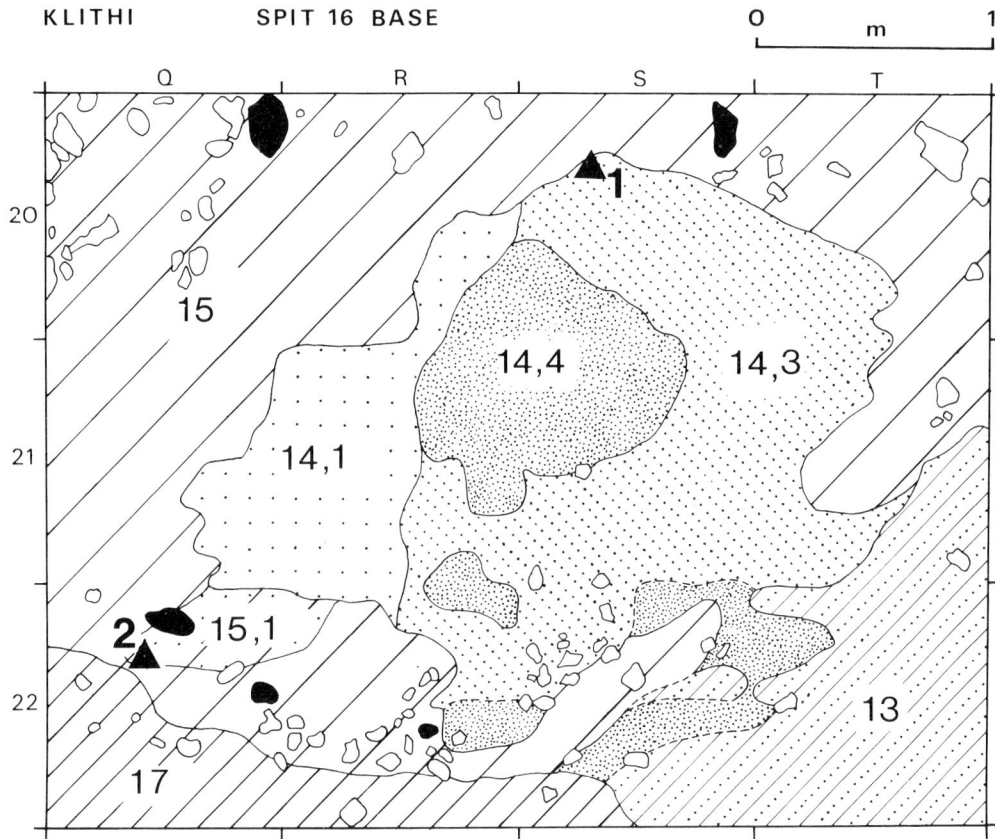

Fig. 2 Horizontal distribution of deposits in the back trench at Klithi Layers 13, 15 and 17 are scree deposits. Layers 14, L1; 14, H3; 14, H4; and 15, H1; are ashy deposits. For further details see Bailey et al. in press. Triangles show location of radiocarbon samples: OxA-542 (1) and OxA-502 (2). Filled outlines are river pebbles.

fragment of charcoal sealed underneath a large stone at the surface of the scree, and the late historic date strengthens hypothesis (1) although it does not necessarily eliminate hypothesis (2). The latest date OxA-748 includes bomb carbon and clearly represents material from the use of the shelter in recent decades.

THE KLITHI PALAEOENVIRONMENT: SITE TO OFF-SITE CORRELATIONS

It is axiomatic that any attempt to place archaeological sites in their contemporary environmental setting must take account of changes in topography, sediments, hydrology and vegetation cover since (and during) the time of occupation. Palaeoenvironmental work in the vicinity of Klithi has therefore concentrated so far on two types of sediments: (1) lake deposits which preserve a stratified pollen record of vegetational change; (2) river terraces

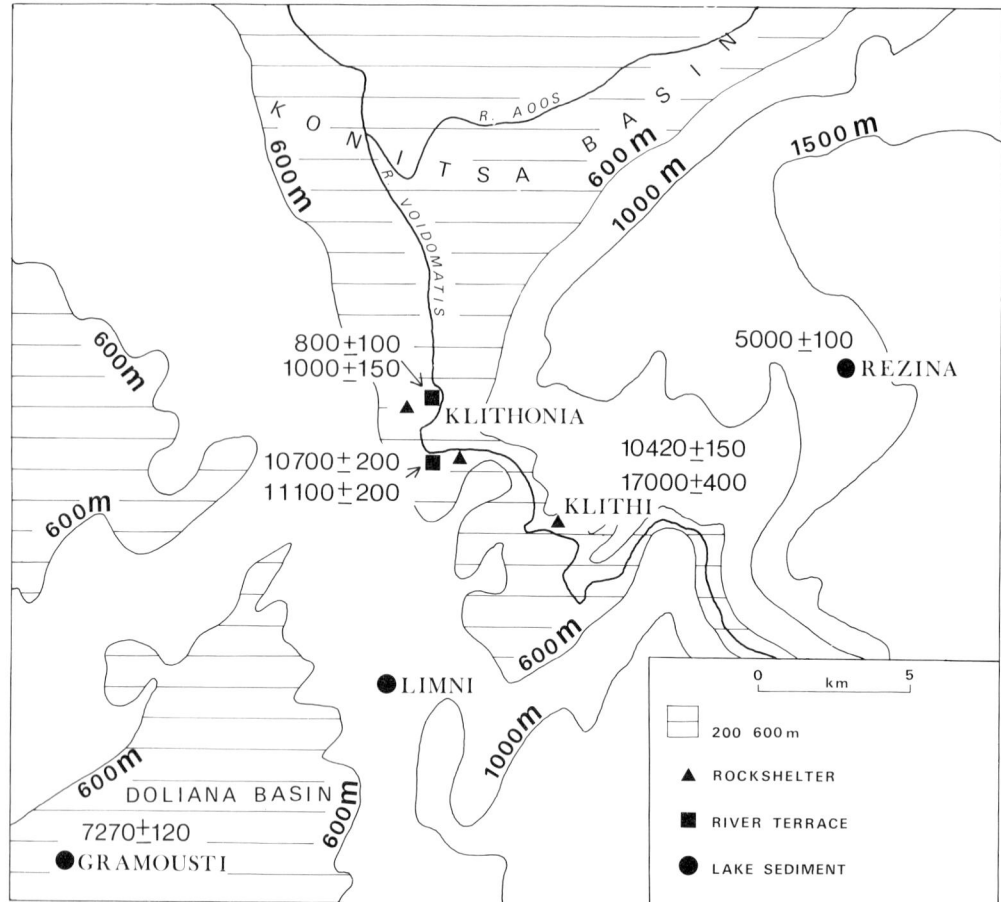

Fig. 3   Location of sites in the vicinity of Klithi

of coarse gravels and silts which indicate changes of sediment cover and hydrology over extensive areas of potential importance to Palaeolithic subsistence (Fig. 3). The former deposits are relatively easy to date, because of generally good organic preservation, but difficult to interpret in terms of the geographical distribution of the vegetation patterns recorded in the pollen core. The latter deposits are relatively easy to map in terms of their geographical extent and potential impact on human settlement, but exceedingly difficult to date because of the rarity or stratigraphic unreliability of organic materials.

   Three small lake basins with good prospects for obtaining a local pollen record have so far been selected for pollen sampling. Two (Gramousti and Limni) are recently drained deposits, and in two cases (Gramousti and Rezina) organic preservation is good. At Gramousti the depth of the borehole has been almost doubled since the first date (OxA-512) was obtained, so that there is a good chance of recovering a continuous record extending from historical times back into the terminal Pleistocene period contemporaneous with the occupation deposits at Klithi. At Rezina it appears likely that lake sediments only started accumulating in the early Holocene, but the results of pollen analysis here will nevertheless provide a useful spatial control on the interpretation of the other sequences.

Two localities in the river gravels flanking the lower reaches of the Voidomatis River have yielded datable materials (Gillespie *et al.* 1985). A hearth in sediments *ca.* 8 m above the present river level is dated to 11,000 BP (OxA-351, OxA-353). The older readings obtained from the humic acids (OxA-352, OxA-372) are attributed to the accumulation of older terrestrial soil over the fluviatile sediments after the river had begun to deepen its channel presumably in the early Holocene. Higher river gravels up to 18 m above local river level are recorded in the gorge close to Klithi. The occupation deposits at Klithi are 28–30 m above present river level, so that a river channel at 18 m would have greatly altered the immediate surroundings of the site and made it more easily accessible. But these high-level deposits are not yet dated. Further out onto the Konitsa Plain OxA-191 and OxA-192 provide a minimum age for the colluvial fill of gullies eroded into older terrace gravels. Observations of these and other sections through Pleistocene and Holocene deposits, and analysis of landforms, backed up by the radiocarbon dates, suggest that the Konitsa Plain in the period 20,000 BP to 10,000 BP would have been an expanse of gravels and sands subject to an intense seasonal flood with little vegetation and no developed soils, in contrast to the fertile conditions which exist today.

KLITHI IN REGIONAL CONTEXT: INTER-SITE CORRELATIONS

In Figure 4 the sequences of the three dated and excavated Epirus rockshelters are compared in a simplified way. These sites lie on a transect from coastal lowland to high altitude environments, and according to existing interpretations (Higgs *et al.* 1967, Bailey *et al.* 1983a) should be considered as complementary and hence contemporaneous elements in a single system of land use which persisted throughout the Upper Palaeolithic sequence. It is apparent, however, that the available dates are consistent with an alternative and equally plausible hypothesis in which the warmer coastal lowlands represent the core area of human settlement throughout the extremes of last glacial climate, while successive use of more inland and upland areas occurred only with the progressive amelioration of climate after the last glacial maximum. There are still many uncertainties in this picture. The Asprochaliko stratigraphy is very complex with only one reliable radiocarbon date available. Kastritsa was not available for occupation before 20,000 BP because of high lake levels, and the apparent increase in scale of activity in the upper deposits might simply represent a lateral shift within the shelter of the zone of most concentrated activity and discard. The concentration of finds in the uppermost levels of Klithi could likewise be a function of lateral variation rather than a time trend between 17,000 BP and 10,000 BP, and we do not yet know what archaeological deposits if any lie at a deeper level. Such an exercise in inter-site correlation is nevertheless useful in highlighting alternative hypotheses and indicating fresh lines of inquiry.

ACKNOWLEDGEMENTS

The fieldwork referred to above was carried out with financial assistance from the British Academy, the British School of Archaeology at Athens, the National Geographic Society, the Society of Antiquaries and the Science and Engineering Research Council. We are indebted for the issue of permits for fieldwork and export of dating samples to: the British

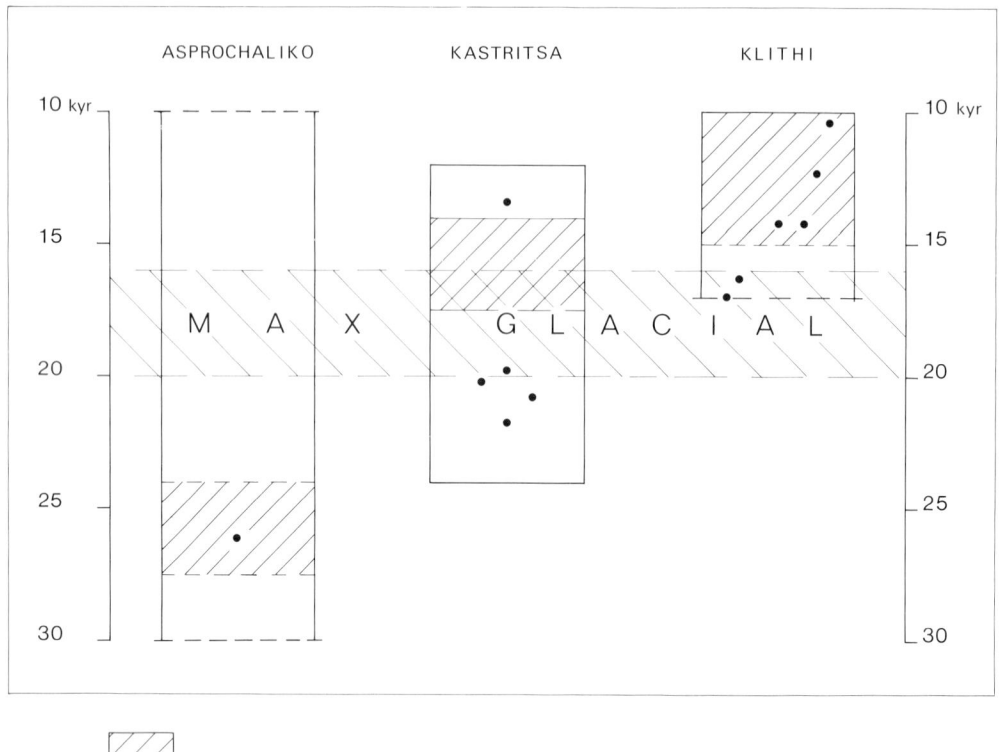

Fig. 4 Chronological relationships between Upper Palaeolithic deposits in Epirus. For Klithi radiocarbon dates see Table 1. For details of dates from the other sites, see Bailey et al. 1983b.

School of Archaeology at Athens; the Ministry of Culture and Science, Directorate of Prehistoric and Classical Antiquities, Athens; the Ephorate of Prehistoric and Classical Antiquities, Ioannina; the Ephorate of Speleology and Palaeoanthropology, Athens; and the Institute of Geological and Mineralogical Research, Athens.

REFERENCES

Bailey, G.N., Carter, P.L., Gamble, C.S. and Higgs, H.P., 1983a, Epirus revisited: seasonality and inter-site variation in the Upper Palaeolithic of north-west Greece, in *Hunter-gatherer economy in prehistory: a European perspective* (ed. G.N. Bailey), pp. 64–78, Cambridge: University Press.

Bailey, G.N., Carter, P.L., Gamble, C.S. and Higgs, H.P., 1983b, Asprochaliko and Kastritsa: further investigations of Palaeolithic settlement and economy in Epirus (north-west Greece), *Proc. Prehist. Soc.* 49, 15–42.

Bailey, G.N., Carter, P.L., Gamble, C.S. and Higgs, H.P. and Roubet, C., 1984, Palaeolithic investigations in Epirus: the results of the first season's excavations at Klithi, 1983, *Annual of the British School of Archaeology at Athens* 79, 7–22.

Bailey, G.N., Gamble,C.S., Higgs, H.P., Roubet, C., Sturdy, D.A. and Webley, D.P., in press,

Palaeolithic investigations at Klithi: preliminary results of the 1984–1985 field seasons, *Annual of the British School of Archaeology at Athens* 81.

Dakaris, S.I., Higgs, E.S. and Hey, R.W., 1964, The climate, environment and industries of Stone Age Greece, part I, *Proc. Prehist. Soc.* 30, 199–244.

Gillespie, R., Gowlett, J.A.J., Hall, E.T., Hedges, R.E.M. and Perry, C., 1985, Radiocarbon dates from the Oxford AMS system: Archaeometry Datelist 2, *Archaeometry* 27, 2, 237–246.

Higgs, E.S. and Vita-Finzi, C., 1966, The climate, environment and industries of Stone Age Greece, part II, *Proc. Prehist. Soc.* 32, 1–29.

Higgs, E.S., Vita-Finzi, C., Harris, D.R. and Fagg, A.E., 1967, The climate, environment and industries of Stone Age Greece, part III, *Proc. Prehist. Soc.* 33, 1–29.

Sordinas, A., 1969, Investigations of the prehistory of Corfu during 1964–1966, *Balkan Studies* 10, 393–424.

RADIOCARBON ACCELERATOR DATES FOR UPPER PALAEOLITHIC SITES IN EUROPEAN U.S.S.R.

Olga Soffer

I. INTRODUCTION

The Upper Palaeolithic record of the European USSR is a rich and varied one replete with open-air sites with mammoth bone dwellings, well preserved features such as storage pits and hearths, and large inventories of lithic and organic remains. One of the main problems besetting this record has been the question of chronology (Klein 1973; Soffer 1985). The sites, until quite recently, were dated on traditional lithic stylistic criteria — a practice marred by serious problems and one which, expectedly, brought forth conflicting interpretations. The 1950s and 1960s saw large-scale efforts to develop regional stratigraphic schemes to date the sites (e.g. Velichko 1961), as well as first efforts to use isotope dating. The resulting radiometric dates obtained until now, both in the USSR and abroad, are few in number and contain some enigmatic inconsistencies (for a complete list of all ^{14}C dates available for Soviet sites see Boriskovskii 1984; for Russian Plain Upper Palaeolithic sites Kurenkova 1980; Soffer 1985). This report offers new dates for 11 Upper Palaeolithic sites in the European USSR. Five sites are dated for the first time (Fig. 1).

2. METHODOLOGY

A salient feature of these sites, one noted by numerous authors, is the almost complete absence of wood charcoal in hearths which instead contain an abundance of burnt bone (Klein 1973; Soffer 1985). The sites do also contain rich inventories of organic remains such as unburnt bone, teeth, and antlers, and all of these media have been used for radiometric dating in the past. Dates reported here were obtained in two cases on burnt bone from hearths (Kamennaya Balka II and Yudinovo), and the remainder on unburnt mammoth teeth from cultural layers. All the unburnt bones were dated from amino acids extracted by the methods given in Gillespie et al. 1984. In the case of Kamennaya Balka II sufficient amino acids remained for the bone to be dated as if unburnt, and then again from its charred residue. (A discussion of dating of charred material and of separate fractions of a single sample is given in Batten et al. 1986.)

3. DATED SITES

(1) Berdyzh

OxA-716: 15,000 ± 250 on unburnt mammoth tooth from the cultural layer found on the eastern bank of Sozh river some 500 m south of the village of Podluzhye (52°50′N, 30°58′E) in Byelorussian SSR. Cultural remains were found in grey-green sandy loam deposits in a ravine at depths from 1.5 to 10 m below present day surface. Kalechitz (1984) argued that this stratum was solifluctory in origin and that organic and lithic inventories were in redeposited secondary context. Budko (1966), Budko and Voznyachuk (1969), and Voznyachuk and Budko (1969) wrote that parts of the cultural layer were still *in situ*. Stylistic lithic criteria led Soviet scholars to see the site as a very early Upper Palaeolithic one (Boriskovskii 1984; Kalechitz 1984). The new accelerator date is considerably younger than the one existing date reported by Leningrad University (Lu-104: 23,430 ± 180) which was also obtained on a mammoth tooth. The discrepancies between the dates may well be a result of the dated medium (see discussion below).

(2) Chulatovo I

OxA-715: 14,700 ± 250 on mammoth tooth from either a severely disturbed (Boriskovskii 1953; Pidoplichko 1947a) or a totally redeposited cultural layer (Velichko 1961) found on the northern bank of the Desna river in a Ukrainian village of Chulatovo (51°50′N, 33°07′E). Soviet researchers used lithic criteria to assign the site to the Magdalenian period (*sensu lato*) (Boriskovskii 1953; Shovkoplyas 1965a, 1967; Voevodskii 1947). This is an agreement with the first radiometric date for the site reported here.

(3) Dobranichevka

OxA-700: 12,700 ± 200 on mammoth tooth from an *in situ* cultural layer found in loess III (Dolukhanov and Pashkevich 1977; Pashkevich and Dubyank 1978). The site is located on a left bank promontory formed by the second terrace of the river Tashanka where it enters the valley of the Supoi river in the Ukrainian village of Dobranichevka (50°02′N, 32°40′E). This first radiometric date for the site is in fairly good accordance with its chronological assignment on stylistic (Shovkoplyas 1976; Gladkih 1973) and geologic criteria (Shovkoplyas *et al.* 1981).

(4) Gontsy

OxA-717: 14,600 ± 200 on mammoth tooth from an in situ cultural layer found in banded loam and sandy deposits on the southern promontory of the second terrace of the Udai river near the Ukrainian village of Gontsy (50°10′N, 32°49′E). The accelerator date reported above, while somewhat older than one obtained on burnt bone (QC-898: 13,400 ± 185, Soffer 1985) is in fairly good accord with it and with chronological assignment of the site based on lithic and stratigraphic criteria (Boriskovskii 1984; Pidoplichko 1969; Gladkih 1973; Shovkoplyas 1965a; Gromov 1948).

(5) Kamennaya Balka II

OxA-699: 10,900 ± 400 on amino acids from a partly-burnt bone from a hearth of an *in situ* cultural layer found in soil stratum about 1.0 m below the present day surface.
OxA-778: 13,660 ± 180 carbonaceous residue of charred bone (same specimen). The site is located on the right bank of a deep old ravine which enters the Mertvii Donets river in the village of Nedvigovka (47°16′N; 39°21′E) in R.S.F.S.R. (Gvozdover 1967; Leonova 1985). At the time of occupation the site was located in the southern steppe zone of the Russian Plain. Its lithic inventory has been assigned to the Emeritan culture (Boriskovskii 1984; Gvozdover 1967; Leonova 1985).

   The Oxford laboratory comments that the discrepancy between the two dates is unusual, and reflects the difficult nature of the sample material. The burning may have destroyed

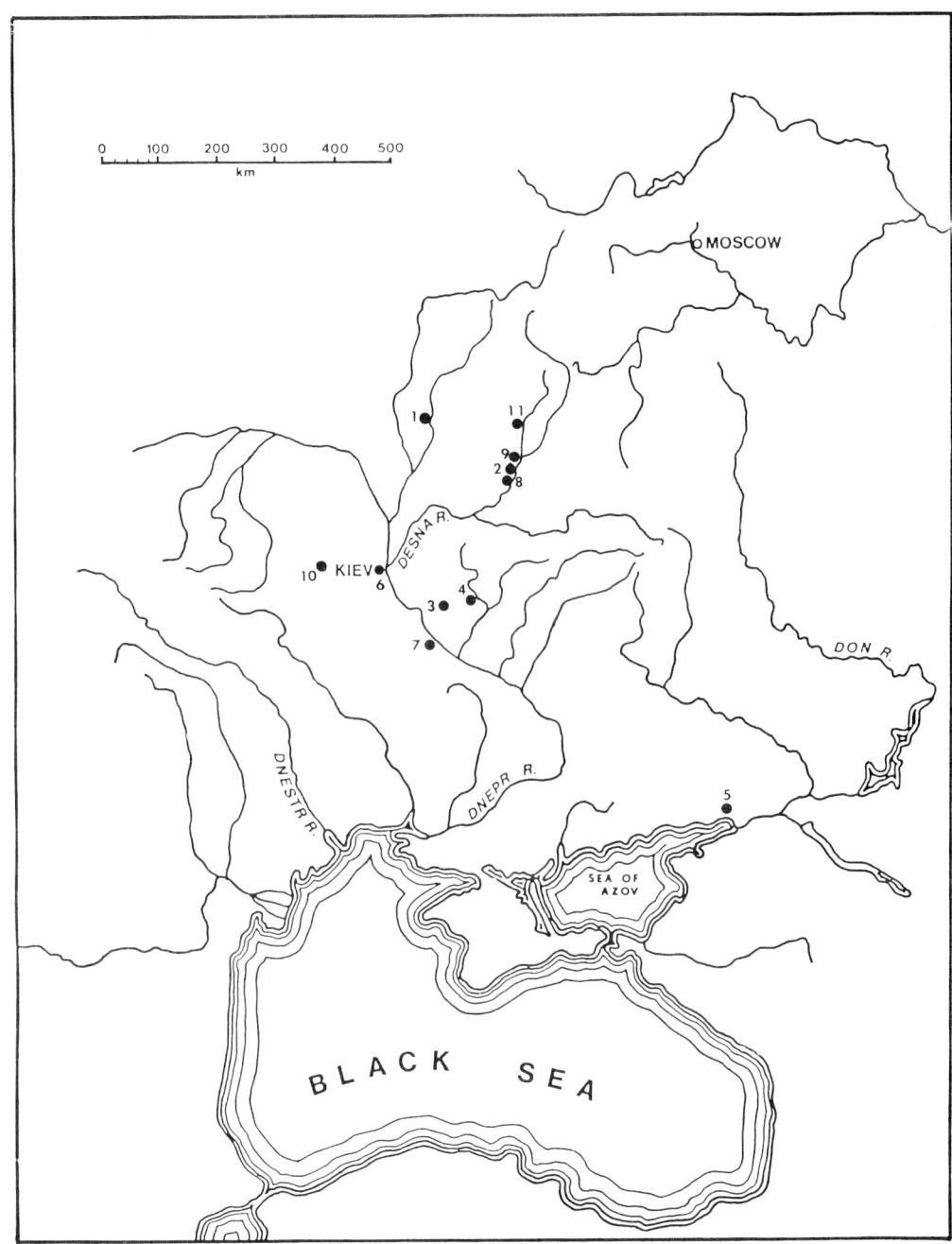

Fig. 1 Upper Palaeolithic sites on the Russian Plain with new accelerator radiocarbon dates:

1 Berdyzh   4 Gontsy   7 Mezhirich   10 Radomyshl'
2 Chulatovo I   5 Kamennaya Balka II   8 Mezin   11 Yudinovo
3 Dobranichevka   6 Kirillovskaya   9 Novgorod-Severskii

most of the original amino acids, and consequently the older date from the carbonaceous residue seems preferable.

(6) Kirillovskaya

OxA-718: $19,200 \pm 350$ on mammoth tooth from a site in Kiev (50°26'N, 30°31'E) and completely excavated at the end of the last century. The cultural layer, as best could be reconstructed from the original excavation notes, lay at depths from 13 to 22 m below the surface, was predominantly *in situ*, and probably contained a number of such features as mammoth bone dwellings, storage pits, hearths, etc. (Boriskovskii 1953; Gromov 1948; Pidoplichko 1969; Soffer 1985). This first radiometric date is in fairly good accord with geologic and stylistic chronological assignment of the site to the later part of the Upper Palaeolithic.

(7) Mezhirich

OxA-709: $12,900 \pm 200$ on mammoth tooth from dwelling 1 and OxA-712: $14,400 \pm 250$ on mammoth tooth from dwelling 2 on an *in situ* cultural layer. The site is located in the Ukraine at the confluence of Ros and Rosava rivers in the village of the same name (49°38'N, 31°24'E). The first of the accelerator dates reported here is the youngest obtained for the site; the second is in good accordance with previously reported GIN and QC dates (see Boriskovskii 1984; Kurenkova 1980; Soffer 1985).

(8) Mezin

OxA-719: $15,100 \pm 200$ on mammoth tooth from the 1953 excavations of dwelling 1. The samples come from an *in situ* cultural layer found in colluvial loess deposits laminated with sandy and clayey layers (Pidoplichko 1969). The accelerator date is considerably younger than ^{14}C dates obtained by Soviet laboratories on shell and mammoth teeth (Boriskovskii 1984; Soffer 1985). The accelerator date, however, accords well with the stylistic assignment of the site by Shovkoplyas (1965a) to the Magdalenian period (*sensu lato*), and with Velichko's (1961) stratigraphic dating of the site to the late Valdai (after the glacial maximum).

(9) Novgorod-Severskii

OxA-698: $19,800 \pm 350$ on mammoth tooth from cultural remains found in collapsed rock shelter in the Ukrainian city of Novgorod-Severskii (52°00'N, 33°16'E). Lithic and organic inventories came from either a quite disturbed (Gromov 1948; Pidoplichko 1947b) or a fully redeposited (Velichko 1961) layer. This first radiometric date for the site appears somewhat old. It is, however, in agreement with a fairly early Upper Palaeolithic dating of the site on lithic stylistic criteria (Boriskovskii 1953; Gladkih 1973, pers. comm. 1977; Shovkoplyas 1965a). Velichko (1961) indicated that the geology and geomorphology of this locality were poorly understood and the site has not been dated stratigraphically.

(10) Rudomyshl'

OxA-697: $19,000 \pm 300$ on mammoth tooth from an *in situ* cultural layer located on the right bank of the Teterev river near the Ukrainian city of Radomyshl' (50°33', 29°14'E). Cultural remains were found in loam some 60–80 cm below the surface in terminal moraine deposits of the penultimate glaciation (Soffer 1985). Shovkolyas (1964, 1965b) used lithic stylistic criteria to classify the site as the earliest ('Aurignacian') Upper Palaeolithic occupations on the Russian Plain. The geology and geography of the site have been investigated only perfunctorily and no chronological statigraphic assignment attempted (Soffer 1985).

(11) Yudinovo

OxA-695: 13,300 ± 200 on humic extract from burnt bone and OxA-696: 12,300 ± 200 on carbonaceous residue from burnt bone collected in 1983 from a hearth of an *in situ* cultural layer. The site is located in a village of the same name (52°40'N, 33°17'E) in R.S.F.S.R. It sits on the northern bank of the Sudost river on a promontory formed by the river's terrace and a buried ravine. A single cultural layer is in alluvial loess-like sandy loam at depths from 2.35 to 2.75 m below the present day surface (Soffer 1985). The accelerator date obtained on humic extract is in good accord with other radiometric dates reported for the site (Boriskovskii 1984; Soffer 1985). The date from carbonaceous residue is the youngest obtained for the site to date.

4. DISCUSSION

Two serious problems face researchers interested in establishing radiometric chronologies for Upper Palaeolithic sites in the European USSR. One of these is the absence of burnt wood at many of the sites. This necessitates obtaining dates on other organic remains (burnt and unburnt bone, teeth, shell, etc.) which are less desirable as the dating medium. Unburnt bone is a problematic medium for dating these sites because a strong possibility exists that Late Pleistocene hunter-gatherers on the Russian Plain collected bone, especially skeletal remains of such large sized species as mammoths and reindeer, for use as raw materials for tool manufacturing, as fuel for hearths, and for the construction of dwellings (Soffer 1985). Their cold weather settlements may have been preferentially located in those parts of river valleys which held large accumulations of fluvially deposited skeletal remains ('mammoth bone cemeteries' in permafrost). If radiocarbon dates are obtained on unburnt skeletal remains, what is being dated is the time of death of the individual animal and possibly not the time when its bones were used — the two events may have been separated by hundreds if not thousands of years. A second problem contributing to ambiguity of radiocarbon dating of mammoth teeth may come from a possibly greater susceptibility of teeth to contamination by both older as well as younger materials. This should not however be the case for amino acids extracted from well preserved ivory or dentine. Radiometric dates obtained from mammoth teeth for the sites on the Russian Plain, in general, show a much greater range of variation than do those obtained for the same sites on burnt bone (Soffer 1985). Since the observed variation includes dates both older and younger than those obtained on burnt bone, it cannot be explained only by invoking the 'collecting of bone' argument offered above. Sample mobility and/or chemical contamination must also be taken into account. While dates on teeth, by and large, do appear to be older than those on burnt bone, the existence of younger tooth dates suggests that contamination must also be considered as a contributing factor.

Unfortunately there are no easy and fast solutions to the problems discussed above. For many sites on the Russian Plain, especially those excavated in the last part of the 19th or early 20th century (e.g. Kirillovskaya, Chulatovo I), no organic remains other than occasional unburnt bones and teeth are available for dating today. Postulated collection of bone during the Late Pleistocene strongly suggests that in the future radiometric dating be done on unburnt bones of taxa other than mammoths and reindeer — two prime Upper Palaeolithic 'collectibles'. Given these problems, we suggest that (1) those dates which represent first dates for a particular site be accepted more tentatively than those which can be incorporated into extant sequences, and (2) dates obtained on burnt bone or on bone of taxa other than mammoths and reindeer may be more reliable.

Lastly, it is significant to note that three of the accelerator dates reported here (OxA-718 for Kirillovskaya, OxA-697 for Radomyshl', and OxA-698 for Novgorod-Severskii) indicate occupation of these sites at or near the glacial maximum — a period noted for a dearth of sites in Eastern Central, and Western Europe (Hahn 1986; Soffer 1985).

ACKNOWLEDGEMENTS

O. Soffer's research in the Soviet Union in 1983 when the samples were collected, took place under the auspices of the exchange of scholars program between the National Academy of Sciences of the U.S. and the Academy of Sciences of the U.S.S.R. It was also supported by grants from The Explorer's Club and the National Geographic Committee for Research and Exploration.

REFERENCES

Batten, R., Gillespie, R., Gowlett, J.A.J. and Hedges, R.E.M., 1986, The AMS dating of separate fractions in archaeology. *Proc. of the 12th Int. Radiocarbon Conf., Trondheim, Norway, 1985, Radiocarbon* 28, 2A and B.

Boriskovskii, P.I., 1953, Paleolit Ukraini, *Materiali i Issledovaniya po Arkheologii SSR* 40, Moscow-Leningrad: Izdatel'stovo An SSSR.

Boriskovskii, P.I., 1984, (ed.) *Paleolit SSSR*, Moscow: Nauka.

Budko, V.D., 1966, Verhnii Paleolit severo-zapada Russkoi ravnini, in *Byelorusskiye Drevnosti*, pp. 6–47, Minsk: AN BSSR.

Budko, V.D. and Voznyachuk, L.M., 1969, Nekotoriye rezul'tati raskopok Berdyzhskoi stoyanki v 1969 godu, in *Tezisi dokladov k konferentsii po arkheologii Byelorussii*, pp. 5–7, Minsk: Institut Istroii An BSSR.

Dolukhanov, P.M. and Pashkevich, G.A., 1977, Paleogeografichiskiye rubezhi verhnego Pleistotsena — Golotsena i razvit'ye khozyaistvennih tipov na yugo-vostoke Evropy, in *Paleogeografiya Drevnego Cheloveka*, (eds. I.K. Ivanova and N.D. Praslov), pp. 134–146, Moscow: Nauka.

Gillespie, R., Hedges, R.E.M. and Wand, J.O., 1984, Radiocarbon dating of bone by accelerator mass spectrometry. *J. Archaeol. Sci* 11, 1, 165–170.

Gladkih, M.I., 1973, *Pozdnii Paleolit Lesostepnogo Pridneproviya*, Avtoreferat kandikatskoi dissertatsii na soiskaniye uchenoi stepeni kandidata istoricheskih nauk, Leningradskoye Odteleniye Instituta Arkheologii AN SSSR.

Gromov, V.I., 1948, Paleontologicheskiye i arkheologicheskiye obosnovaniya stratigrafii kontinental'nih otlozhenii Chetvertichnogo perioda na territorii SSR, *Trudy Instituta Geologicheskih Nauk* Vyp. 64, Geologicheskaya Seriya No. 17, Moscow-Leningrad: AN SSSR.

Gvozdover, M.D., 1967, O kul'turnoi prinadlezhnosti Pozdnepaleoliticheskih pamyatnikov nizhnego Dona, *Voprosy Antropologii* vyp. 27, pp. 82–101.

Hahn, J., 1986, Aurignacian and Gravettian settlement patterns in Central Europe, in *Regional perspectives on Pleistocene prehistory of the Old World*, (ed. O. Soffer, New York: Plenum (in press).

Kalechitz, E.G., 1984, *Pervonachal'noye zaseleniye territorii Byelorussii*, Minsk: Nauka i Tekhnika.

Klein, R.G., 1973, *Ice-Age hunters of the Ukraine*, Chicago: University of Chicago Press.

Kurenkova, E.I., 1980, *Radiouglerodnaya khronologiya i paleogeografiya Pozdnepaleoliticheskih stoyanok vernego Pridneprov'ya*, Avtoreferat dissertatsii na soiskaniye uchenoi stepeni kandidata geograficheskih nauk, Moscow: Institut Geografii An SSSR.

Leonova, N.B., 1977, *Zakonomernosti raspredeleniya kremnevogo inventarya na Verhnepaleoliticheskih stoyankah i otrazheniye v nih spetsifiki Paleoliticheskih poselenii*, Avtoreferat dissertatsii na soiskaniye uchenoi stepeni kandidata istoricheskih nauk, Leningradskoye Otdeleniye Instituta Arkheologii AN SSSR.

Leonova, N.B., 1985, Planigraficheskoe issledovanie svidedetl'stv utilizatsii okhotnichei dobychi na materialakh Verhnepaleoliticheskoi stoyanki Kamennaya Balka II, *Kratkiye Soobscheniya Instituta Arkheologii* No. 181, pp. 12–17.

Pashkevich, G.A. and Dubnyak, V.A., 1978, Paleogeograficheskaya kharakteristika razreza s. Dobranichevka, in *Ispol'zovanie metodov estestvennih nauk v arkheologii*, (ed. V.F. Gening), pp. 69–85, Kiev: Naukova Dumka.

Pidoplichko, I.G., 1947a, Paleolitichna stoyanka Chulativ I, in *Paleolit i Neolit Ukraini (Vol. 1, Vyp. II)*, pp. 123–153, Kiev: AN USSR.

Pidoplichko, I.G., 1947b, Piznyopaleolitichna stoyanka Novgorod-Sivers'k, in *Paleolit i Neolit Ukraini (Vol. I, Vyp. II)*, pp. 65–106, Kiev: AN USSR.

Pidoplichko, I.G., 1969, *Pozdnepaleoliticheskiye zhilischa iz kostei mamonta na Ukraine*, Kiev: Naukova Dumka.

Shovkoplyas, I.G., 1964, Paleolitichna stoyanka Radomyshl', *Arkheologiya* XVI, pp. 88–102.

Shovkoplyas, I.G., 1965a, *Mezinskaya stoyanka*, Kiev: Naukova Dumka.

Shovkoplyas, I.G., 1965b, Radomyshl'skaya — pamyatnik nachal'noi pory pozdnego Paleolita, in *Stratigrafiya i periodizatsiya Paleolita Vostochnoi i Tsentral'noi Evropy*, (ed. O.N. Bader, I.K. Ivanona, and A.a. Velichko), pp. 104–116, Moscow: AN SSSR.

Shovkoplyas, I.G., 1967, Novaya pozdnepleoliticheskaya stoyanka na Chernigovschine, *Arkheologicheskiye otkritiya 1966 goda*, pp. 187–189, Moscow: Nauka.

Shovkoplyas, I.G., 1976, Issledovaniya v Dobranichevke na Kievschine, *Arkheologicheskiye otkritiya 1975 goda*, pp. 407–408, Moscow: Nauka.

Shovkoplyas, I.G., Kornietz, N.L. and Pashkevich, G.A., 1981, Dobranichevskaya stoyanka, in *Arkheologiya i paleogeografiya Pozdnego Paleolita Russkoi Ravnini*, (ed. A.A. Velichko), pp. 97–106, Moscow: Nauka.

Soffer, O., 1985, *The Upper Paleolithic of the Central Russian Plain*, Orlando, Florida: Academic.

Velichko, A.A., 1961, *Geologicheskii vozrast Verhnego Paleolita tstentral'nih raionov Russkoi Ravnini*, Moscow: AN SSSR.

Voevodskii, M.V., 1947, Kremyani i kistyani vyroby Paleolitichnoi stoyanki Chulativ I, *Paleolit i Neolit Ukraini*, (Vol. 1, Vyp. XXXI), pp. 40–54.

Voznyachyk, L.M. and Budko, V.D., 1969, O znachenii Berdzyhskoi stoyanki dlya opredeleniya vozrasta rechnikh terras basseina Dnepra, in *Tezisy dokladov k konferentsii po arkheologii Byelorussii*, pp. 7–16, Minsk: AN BSSR.

Section IV
Later Prehistory

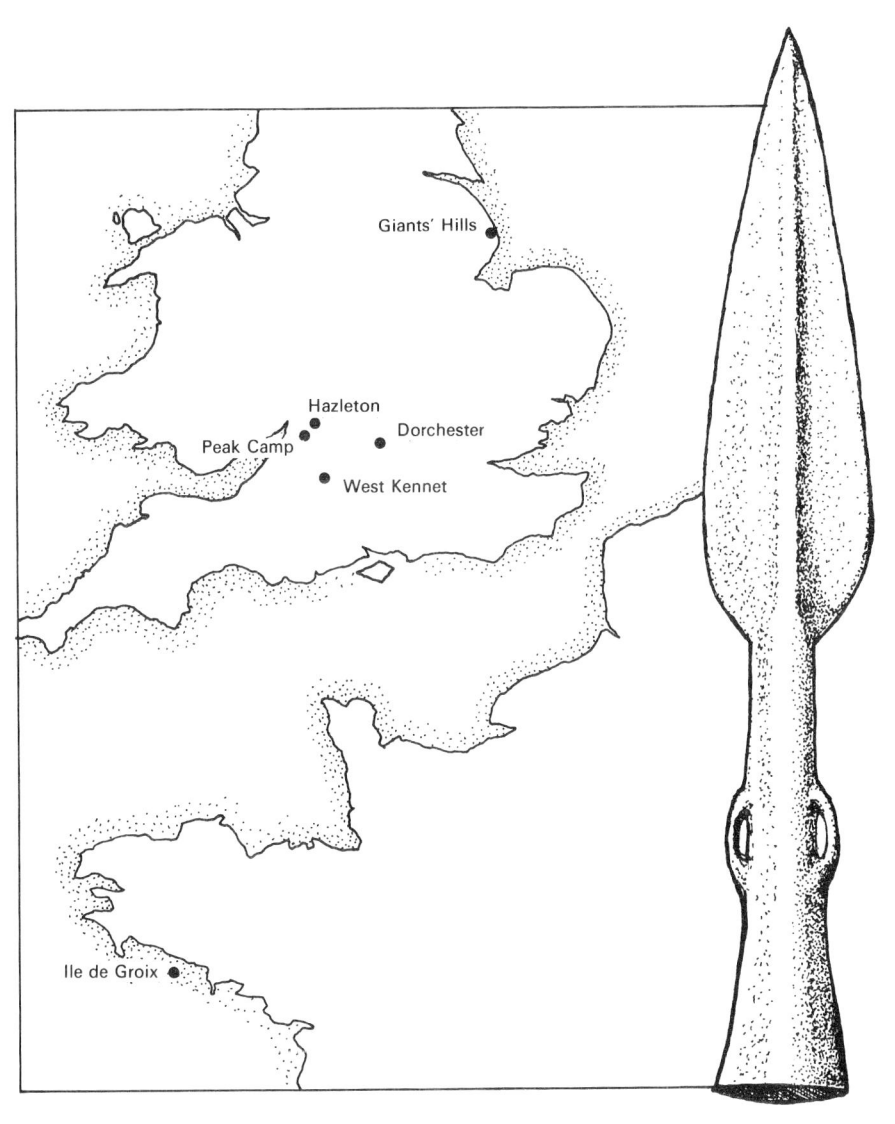

Giants' Hills

Hazleton

Peak Camp

Dorchester

West Kennet

Ile de Groix

Overleaf: Some accelerator dated sites of the Neolithic and Bronze Age in Britain and France; (right) Bronze Age spearhead from Scambleby, Lincs (from a replica, courtesy of Ashmolean Museum). Numbers of Bronze Age spearheads

PROSPECTS FOR DATING NEOLITHIC SITES AND MONUMENTS IN THE COTSWOLDS AND ADJACENT AREAS

T. C. Darvill

INTRODUCTION

Research into Neolithic chronology as a fundamental factor in the interpretation of stratigraphy, artefact association, and inter-site comparison is a relatively underdeveloped aspect of Neolithic studies in Britain. Over 900 dates relating to the Neolithic period are currently available, but most derive from rescue excavations and largely represent attempts to provide absolute dates for the initial construction and use of individual sites and monuments. The potential of the Oxford radiocarbon accelerator as a research tool facilitating the investigation of wider questions of Neolithic chronology was recognised at an early stage in the development of the system, for example in research briefs provided by A.C. Renfrew and T.C. Darvill (see also Darvill 1983). This summary briefly outlines the background to the Neolithic dating project and its main aims, together with a preliminary appraisal of the results from one of the sites selected as a case study: The Peak Camp, Gloucestershire. Throughout the project emphasis has been placed on what is conventionally referred to as the early and middle Neolithic periods. It may be noted that the British Museum is engaged in a dating project concerned with the later Neolithic/Beaker period (Kinnes *et al.* 1983).

HISTORICAL PERSPECTIVE

The British Neolithic underwent an especially radical re-definition during the first radiocarbon revolution. The effect was exacerbated by the fact that the publication of Piggott's (1954) seminal analysis of the period, based on pre-radiocarbon chronologies, was coincident with the determination of the first clutch of dates.

In retrospect, much early use of radiocarbon dates seems infelicitous and haphazard. Bulked samples from contexts widely scattered round a site, single determinations for complex stratigraphic sequences, loose associations between samples and artefacts or fixed horizons, and so on, were common. For these reasons many early determinations can only be used with caution and great attention to what is actually dated.

What has emerged over the past three decades is a broad chronological framework for the Neolithic period which allows the main types of monument to be accommodated within the overall sequence of development, and at least a generalised pattern of change can be established for artefact types (Smith 1974). Naturally there are still some classes of monument which are not yet securely tied down, for example burnt mounds and cursus, but this will no doubt be rectified in due course as more sites are investigated.

Alongside the development of this chronological framework much has been learnt from excavation and fieldwork about the nature of the Neolithic evidence itself. Three important characteristics have been clarified. First, considerable regional variation is evident, especially in the form and sequence of monuments, the types of pottery used and the continuity of settlement. Second, the stratigraphy of some monuments, particularly causewayed enclosures and monumental tombs, suggest long periods of use. Third, some areas of the country have very dense concentrations of Neolithic sites including familiar monuments such as long barrows and enclosures, as well as open sites and industrial sites.

CHRONOLOGICAL DIMENSIONS

Many specific questions connected with Neolithic chronologies could be cited as a result of recent work, but most may conveniently be subsumed under one or other of the following three fields of investigation, or chronological dimensions:

(i)	Site chronologies	The duration of stratigraphically definable phases or episodes of activity (or lack of it), contemporaneity of features within the site, and the overall length of site use.
(ii)	Landscape chronologies	The contemporaneity of individual sites, activity areas, and reconstructable environmental patterns within a defined landscape area. Such an area might well include several contrasting ecozones.
(iii)	Neolithic succession	The national sequence of events, fashions, and monument types. Regional variation in pattern and relationships with other areas of Europe.

In the past much emphasis has been placed on the third and most general level of analysis – the Neolithic succession. Attention focused on dating the start of the period and the order of events within it. In practice, the three proposed levels are hierarchical in the sense that well documented site chronologies will lead to an understanding of the landscape at various times, and this in turn will enable a still broader picture to be glimpsed.

The aim of the Neolithic dating project is to look at all three dimensions by starting with site chronologies and working towards the more general analysis.

THE EVIDENCE AND THE EXPECTATIONS

By comparison with say Iron Age sites, the quantity of artefacts and ecofacts recovered from most British Neolithic sites is small. Charcoal, because of its antiquity is often finely comminuted, and few contexts yield large quantities of animal bone. Human bone, the main class of material recovered from burial monuments, is subject to many demands for specialist study in addition to dating, and of course there are moral pressures not to destroy skeletons in the name of science. For these reasons small sample dating is desirable.

Relying on the vagaries of rescue excavation to provide appropriate information for research in the field of chronology is unrealistic. Individual sites may be accessible in this way, but in order to advance beyond the site based stage other sources of information are needed. One such source is research excavation which provides controls on rescue work and helps to fill out the information available about a given area of landscape. Another important source of evidence is material from old excavations, often 19th or early 20th century work. Well provenanced animal and human bones recovered in the days before radiocarbon dating was available are particularly useful.

Most important to a successful dating programme is balancing the expectations with what can really be achieved. Duplicating samples allows a statistically more acceptable approximation of the true date of specific episodes or events, for example dating several bones from an horizon in a ditch fill may overcome some of the problems of residuality, while dating several bones from one skeleton increases the probability of getting close to its true date. At present, however, it is unrealistic to expect to separate events in any given stratigraphic sequence which are less than about 200 years apart. Thus only fairly widely spaced events in a given sequence can be recognised, although it may be found that the ends of an otherwise overlapping sequence of dates relating to a long series of events can be distinguished. Because of this, assessing the contemporaneity of sites in a landscape means assessing the overlap in the spread of dates (or of grouped dates) for events at each of the sites. In cases where events cannot be separated it has to be accepted that they took place within the period indicated by the spread of dates — within the resolution of the dating technique.

Obtaining groups of datable material from widely scattered sites will only duplicate the kind of information already available. To begin to exploit the hierarchy of dating dimensions already discussed attention needs to be focused on a relatively small area with plenty of material available for analysis. Such a study is underway in Sussex (Bedwin 1981, p. 86). By the happy coincidence of rescue excavation (Saville 1984 and this volume), research work (Darvill 1984) and the presence of material from earlier investigations (see Crawford 1925), the Cotswolds was found to provide a useful area in which to begin work for the accelerator based project. Dates have so far been determined for two sites — Hazleton (Saville this volume) and The Peak Camp.

THE PEAK CAMP, GLOUCESTERSHIRE — A CASE STUDY

The Peak Camp is situated on a promontory of the Cotswold escarpment overlooking the Severn Vale in the parish of Cowley, Gloucestershire (SO 924 150). The presence of earthworks on the hilltop has been known for many years (Aubrey 1665–1695, M41 in Fowles 1982) but it was not until the excavations of 1980 and 1981 that a Neolithic date for the visible features was confirmed. Two trenches were investigated, Area I through the line of the earthworks to establish their nature and stratigraphy and Area II at the western extremity of the site where erosion posed a threat to preserved deposits. Two interim reports have been published (Darvill 1981; 1982) and post-excavation work is well advanced towards final publication.

The rampart section (Area I) revealed a pattern of ditch recutting now familiar from a number af Neolithic enclosures. In the case of The Peak Camp, recutting was accompanied, in the areas investigated at least, by a migration of the medial line of the ditch, thus providing greater stratigraphic resolution of the sequence of ditch cuts. In all, 4 phases of recutting were recognised. The first phase was a square cut and relatively shallow ditch and no finds of any sort were recovered from the fill. The ditches of phases II and III were rather deeper and contained quantities of flintwork, animal bone and pottery. The phase IV ditch was the smallest of all the phases represented, but contained the most finds.

At present 6 dates are available from the rampart section in area I, 5 from phase II and 1 from phase IV. These are plotted out at one standard deviation on Figure 1. Dates for the internal features investigated in area II are still awaited.

The five samples relating to phase II show a tight cluster with good overlap at 2 places of standard deviation. Samples OxA-445 (4670 ± 90 BP) from bone and OxA-446 (4810 ± 90 BP) from a tooth both derived from a single mandible. The difference in the means of about 1.5 standard deviations is quite in accordance with statistical expectation. The pooled mean for the four dates on samples from layers 19 and 20 in the middle of the ditch fill is 4700 ± 45 BP (following Ward and Wilson 1978). The event dated is the filling of the ditch, and the position of material sampled in charcoal spreads above a massive deposit of rubble which probably represents the collapse of the bank suggests a fairly advanced stage of filling. Thus the pooled date represents the end of the archaeologically definable phase II rampart construction and use.

The single date of 4290 ± 80 BP so far available for Phase IV is statistically distinguishable from those for Phase II and suggests that use of the site continued, or was re-established, in the late Neolithic. More dates will be needed to substantiate this apparent difference, but it is worth noting at this stage that no late Neolithic pottery (Peterborough series or Grooved Ware) was found during the excavations. Similar final recuts have been noted at other enclosures, for example Hambledon Hill, Dorset (Mercer 1980, pp. 35–6) and Briar Hill, Northamptonshire (Bamford 1985, p. 39).

It is too early to draw firm conclusions from The Peak Camp series of dates but they do provide encouragement that it will be possible to look for chronological differences between phases. Clearly a long period of use is represented, perhaps 6 or 7 centuries, although whether the site was continuously or intermittently occupied is not yet clear. It is, however, worth noting that The Peak Camp phase II dates fall squarely in the centre of the spread of dates so far available from burials in the Hazleton North long barrow.

PROPOSED WORK

Building up sufficiently detailed and numerous site based chronologies to allow the development of a landscape sequence cannot be achieved overnight. Five or six sites in the central Cotswolds near to Hazleton and The Peak Camp can provide material for dating, mostly human bone from 19th century excavations, and with these results to hand the picture of settlement in this area will be better understood.

Further afield the Avebury area of North Wiltshire could provide another useful set of samples. Sites in this area are frequently used to illustrate or elucidate early and middle Neolithic patterns of change and land-use, but in fact the evidence is at present poorly dated. Three determinations have already been made on bones from West Kennet long barrow and these are broadly similar to those for the burials at Hazleton. Material from other barrows and enclosure sites in the area could be used to date significant contexts, and when complete will provide a useful comparison for the Cotswold and Sussex studies.

REFERENCES

Bamford, H., 1985, *Briar Hill: Excavation 1974–1978*, Northampton Development Corporation.
Bedwin, O., 1981, Excavations at the Neolithic enclosure on Bury Hill, Houghton West Sussex 1979, *Proc. Prehist. Soc.* 47, 69–86.

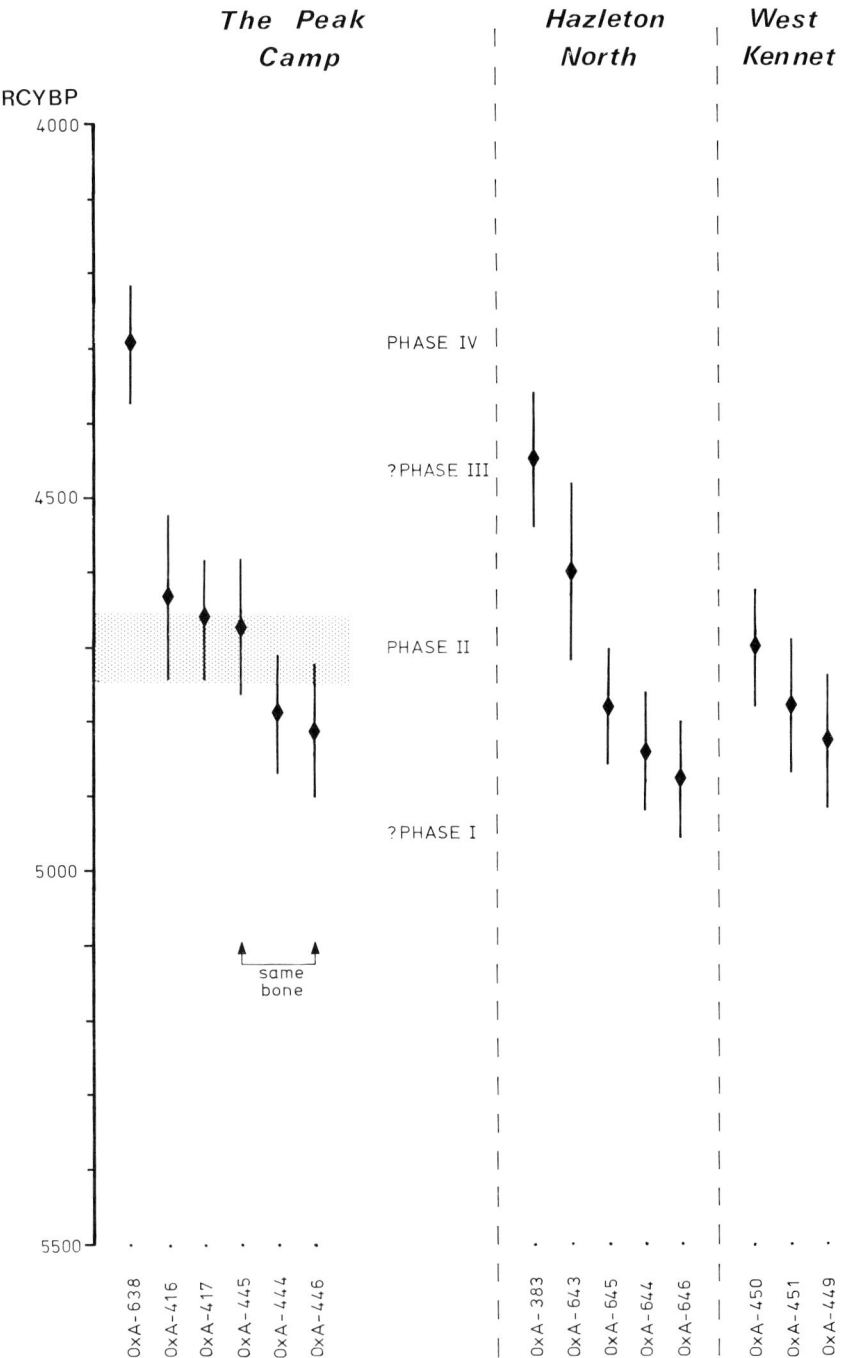

Fig. 1 Plot of radiocarbon dates from The Peak Camp, Gloucestershire, Hazleton long barrow, Gloucestershire and West Kennet long barrow, Wiltshire. The shaded area indicates the pooled means of OxA-416, 417, 445 and 446 at one place of standard deviation.

Crawford, O.G.S., 1925, *The long barrows of the Cotswolds*, Gloucester: John Bellows.

Darvill, T.C., 1981, Excavations at The Peak Camp, Cowley — an interim note, *Glevensis* 15, 52–56.

Darvill, T.C., 1982, Excavations at The Peak Camp, Cowley, Gloucestershire, *Glevensis* 16, 20–25.

Darvill, T.C., 1983, The Neolithic of Wales and the mid-west of England: a systemic analysis of social change through the application of action theory, Unpublished Ph.D. thesis, University of Southampton.

Darvill, T.C., 1984, Neolithic Gloucestershire, in *Archaeology in Gloucestershire — From the earliest hunters to the industrial age* (ed. A. Saville), pp. 80–112, Cheltenham: Cheltenham Museum and Art Gallery & Bristol and Gloucestershire Archaeological Society.

Fowles, J., 1982, John Aubrey's *Monumenta Britannica*, (2 vols), Sherbourne: Dorset Publishing Company.

Kinnes, I., Gibson, A., Burleigh, R., 1983, A dating programme for British beakers, *Antiquity 57*, 218–9.

Mercer, R., 1980, *Hambledon Hill: A Neolithic landscape*, Edinburgh: The University Press.

Piggott, S., 1954, *The Neolithic cultures of the British Isles*, Cambridge: The University Press.

Saville, A., 1984, Preliminary report on the excavation of a Cotswold-Severn tomb at Hazleton, Gloucestershire, *Antiq. J.* 64, 10–24.

Smith, I.F., 1974, The Neolithic, in *British Prehistory: a new outline*, (ed. A.C. Renfrew), pp. 100–136, London: Duckworth.

Ward, G.K. and Wilson, S.R., 1978, Procedures for comparing and combining radiocarbon age determinations: a critique, *Archaeometry* 20, 1, 19–31.

RADIOCARBON DATES FOR THE GIANTS' HILLS 2 LONG BARROW, SKENDLEBY, LINCOLNSHIRE

J. G. Evans and D. D. A. Simpson

Giants' Hills 2 is one of a pair of Neolithic long barrows in the parish of Skendleby at the southern end of the Chalk uplands of the Lincolnshire Wolds (NGR: TF(53)429709; Lat. 53°12'40"N, Long. 0°8'30"E). It was excavated in 1975 and 1976 by the authors on behalf of the Department of the Environment (now the Historic Buildings and Monuments Commission for England). The other barrow of the pair, Giants' Hills 1, was excavated in 1933 and 1934 by C.W. Phillips (1936). In this paper the radiocarbon dates for Giants' Hills 2 are presented and discussed.

The barrow consisted of a mound of earth and chalk rubble surrounded by a quarry ditch (Fig. 1). At the west end an earlier ditch had been backfilled and the site extended, but no dates are available for this feature. Beneath the mound were three features which appeared on stratigraphical grounds to belong to a distinct pre-mound phase. These were a façade, four posts, and a mortuary area.

The remains of the façade were situated towards the east end of the site and comprised a trench with the casts of timber posts. These had been burnt in an oxidising environment as was shown by the fine nature of the charcoal. The burning had taken place prior to the construction of the barrow which overlay the façade trench infilling.

No datable material was available for the four backfilled postholes on the north side of the barrow.

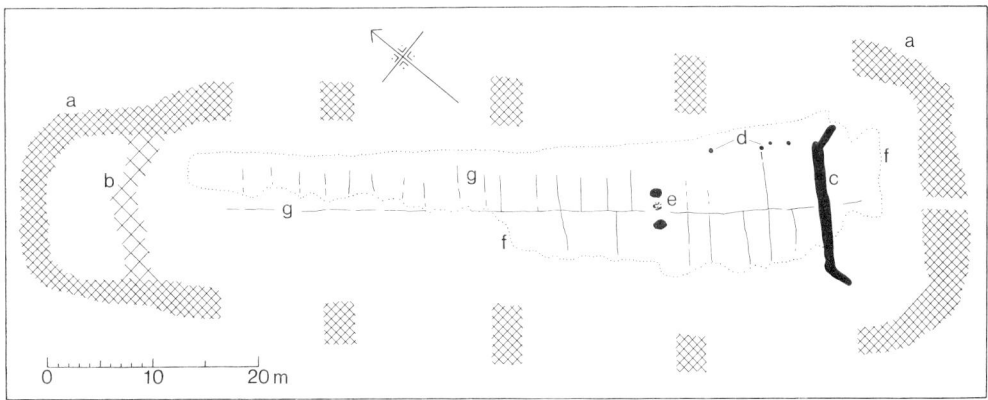

Fig 1. *Giants' Hills 2, Skendleby. Simplified barrow plan. a = ditch; b = backfilled ditch; c = façade trench; d = postholes; e = mortuary area (burial deposit and post-pits); f = preserved edge of mound; g = fences.*

The mortuary area consisted of two post-pits, each probably once containing a half tree-trunk, flanking a small cluster of human bones (Figs. 2 and 3). The posts had not been destroyed by burning but had rotted *in situ*, the casts consisting of clean chalk rubble contiguous with the material of the overlying mound. However, small quantities of charcoal were associated with the north post-pit, possibly deriving from the surface of the post when it was burnt during felling or preparation.

The human skeletal material was fragmentary, consisting of a small concentration of totally disarticulated bones. There were fragments of two skulls and a number of long bones, together with a few ribs and vertebrae. But small bones were virtually absent and none of the long bones had articular surfaces. This was clearly a secondary collection.

The mound was made up of a series of bays defined by a timber framework which was infilled with chalk and turf. Although there was some evidence for sequential construction over a few years, it was essentially a single period feature. In the mound were several red deer (*Cervus elaphus*) antlers which had probably been used in its construction.

The quarry ditch likewise, with the exception of the backfilled western part, was of single period construction. The infilling showed a succession of ceramic and environmental phases (Fig. 2) with four main units — primary fill (layers 7 and 6), secondary fill (layers 5 and 4), buried soil (layer 3) and tertiary fill (layers 2 and 1). Red deer antler on the ditch bottom and in layers 7 and 6 had probably been used in excavating the ditch. Layers 5 and, especially, 4 contained large quantities of charcoal, animal bones, antler and later Neolithic pottery (Ebbsfleet and Mortlake styles of the Peterborough tradition). This material represents re-use of the site rather than casual occurrences. The buried soil formed from the later Neolithic through to the Roman period. There was Beaker pottery at its base and Bronze Age sherds within it. The tertiary fill was largely a ploughwash of Roman and later origin.

With the exception of the two Harwell measurements which were done in 1975, all dates were obtained in 1985. Seven samples were dated in the Cardiff Laboratory (Department of Plant Science), one in the British Museum Research Laboratory, and four in the Oxford Radiocarbon Accelerator Unit. The reasons for using the accelerator were twofold: (1) The two charcoal samples from the façade were too small for conventional dating. (2) The two human skulls were deemed too valuable to destroy and in any case the amount of material was insufficient for conventional dating.

THE RESULTS

The provenance (Figs. 1, 2 and 3) and composition of the samples dated, their excavation reference numbers, the radiocarbon laboratory numbers and the dates in years BP (before present, uncalibrated with reference to any dendro-chronological calibration curve) are listed in Table 1. The dates may be considered to fall into three chronological groups, although they have been plotted in order of age as suggested by archaeological context (Fig. 4).

Five dates fall at or prior to 5000 BP. These include two from the façade and two from the north post-cast of the mortuary area. The earliest date (OxA-641) may be on very old timber but it was not possible either to identify the wood or to age it. The fifth date in this group (HAR-1869), from antler on the ditch bottom, is out of place by comparison with other dates for the construction of the mound (to which it might be thought that the antler

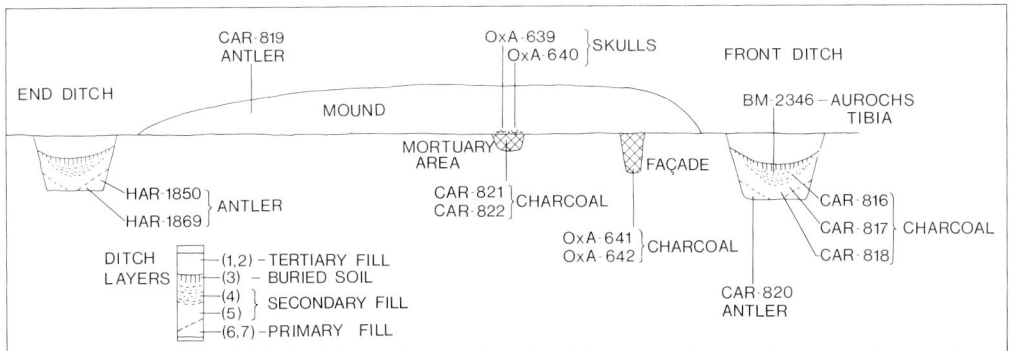

Fig. 2. Giants' Hills 2, Skendleby. Diagrammatic longitudinal section through the barrow to show the provenance of the radiocarbon-dated samples. Note that the earlier backfilled ditch at the west end of the site is not shown.

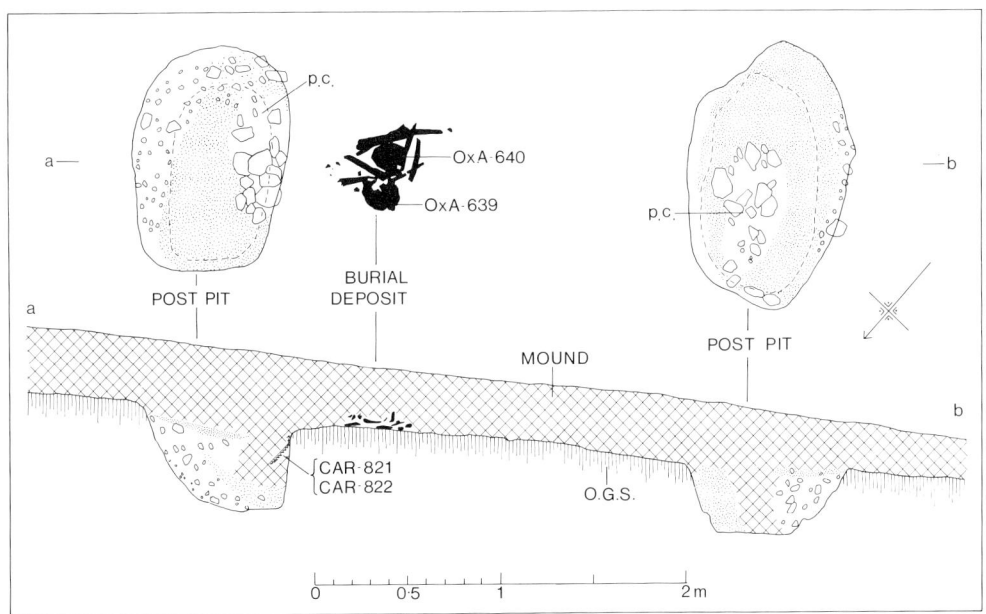

Fig. 3. Giants' Hills 2, Skendleby. Plan (above) and section (below) of the mortuary area. O.G.S. = old ground surface (pre-mound soil); p.c. = post cast. In the section the mound material has been simplified; note that it is contiguous with the post casts. (Note that the post-pit *is the entire hole whereas the* post cast *is the infilled hole left by the decayed post.)*

relates), but two reasons for this discrepancy may be noted. Firstly, the sample was small and, according to the laboratory, 'caused problems'. Secondly, the antler may have been used at a period earlier than mound construction, such as in excavating the façade trench or the earlier ditch at the west end, and thus may have been in a secondary context. This is a general problem with the radiocarbon dating of antler.

In the second age group, five dates fall close together between about 4900 and 4600 BP. These are from antler from the mound, antler from the ditch bottom and primary fill, and the two human skulls. There is substantial overlap with only one date of the earlier group, CAR-822, and it may be noted that this sample was small and gave a measurement with a larger than usual standard deviation.

Four dates spanning the period from about 4500 to 3700 BP come from the secondary fill of the ditch. They relate to the renewed Neolithic activity on the site. The charcoal dates are all from timber having an estimated age of less than twenty years.

We may thus recognise three distinct phases of Neolithic activity as follows (Fig. 4):

Facade and mortuary structure   (5140 ± 80 to 4970 ± 100 BP — excluding (OxA-641)). The four postholes on the north side of the site probably belong to this phase. The façade was destroyed by fire at some time prior to the next phase.

Burial deposition and mound construction   (4840 ± 70 to 4650 ± 80 BP)

Later Neolithic re-use   (4450 ± 70 to 3830 ± 60 BP). It is to this phase that the two radiocarbon dates from Giants' Hills 1 probably belong (Barker *et al.* 1969, 287). They are from antler (unprovenanced) and are: 4410 ± 150 BP (BM-191) and 4320 ± 150 BP (BM-192).

THE ENVIRONMENT

Molluscan, mammalian and charcoal remains show that prior to Neolithic activity there was closed woodland of *Quercus* and *Corylus*, with some *Crataegus* and *Acer*. Clearance to grassland preceeded the first phase of the monument, and there was some cultivation.

Secondary woodland of *Quercus*, *Corylus*, *Fraxinus* and *Taxus* with some *Crataegus* and *Prunus* became established during the period of later Neolithic activity (ditch layer 4). Clearance to grassland took place at a horizon coincident with the Beaker pottery (base of ditch layer 3).

DISCUSSION

The environmental and archaeological sequences are summarised side by side in Fig. 5.

As was the case in parts of the southern English chalklands (Evans 1972), the area of the Lincolnshire Wolds in which Giants' Hills 2 was situated saw a mid-Postglacial vegetation of deciduous woodland. And, as in those southern English areas, there was clearance of this woodland, cultivation of the cleared land, and succession to grassland perhaps as early as 5500 BP.

The monument itself was of two distinct phases. Prior to the building of the mound and prior to the deaths of the individuals whose fragmentary remains were placed beneath it, a free-standing timber façade and a pair of massive split tree-trunk posts stood for several centuries, the posts flanking the area of the future burial. A similar sequence has been

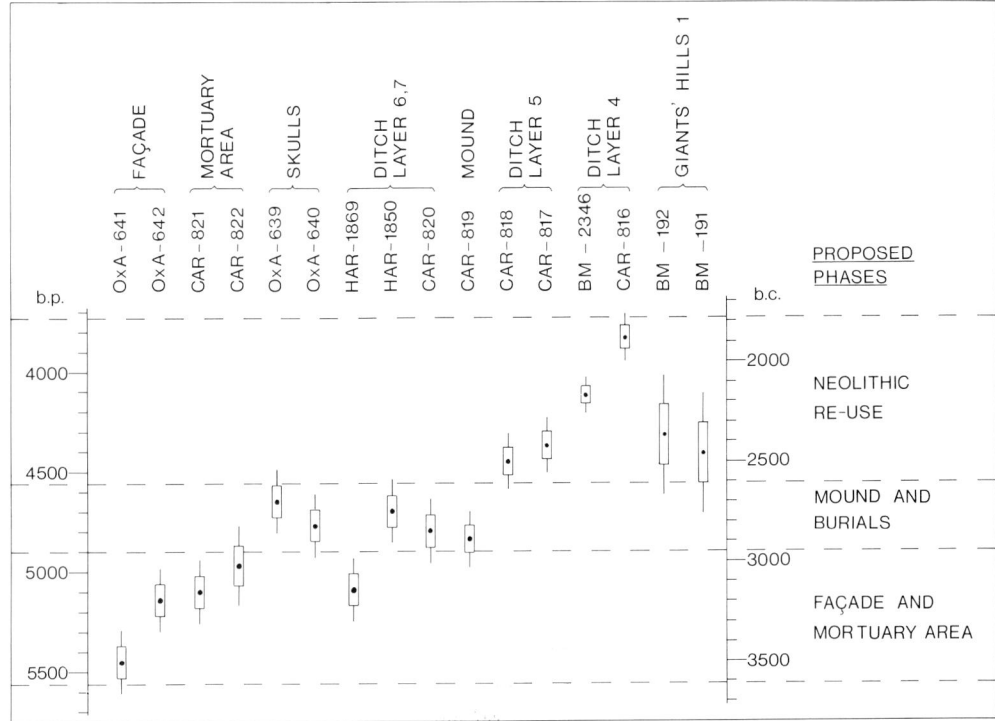

Fig. 4. Radiocarbon dates at one and two standard deviations for Giants' Hills 1 (BM-191, BM-192) and Giants' Hills 2 as issued by the various radiocarbon-dating laboratories.

Table 1

Giant's Hills 2 Long Barrow, Skendleby, Lincolnshire. Radiocarbon dates in years BP as issued by the measuring laboratories. Dates are listed in the order suggested by their archaeological context (prior to radiocarbon dating), and the same order is used in Fig. 4. Numbers in brackets refer to layers.

Lab No.	Material, Reference Number, Context	Date BP
OxA-641	Charcoal 814, façade post	5450 ± 80
OxA-642	*Crataegus* charcoal 815, façade post	5140 ± 80
CAR-821	*Quercus* charcoal 505, north post-pit	5100 ± 80
CAR-822	*Quercus* charcoal 527, north post-pit	4970 ± 100
OxA-639	Human skull 7	4650 ± 80
OxA-640	Human skull 8	4770 ± 80
HAR-1869	*Cervus* antler B 139/752, ditch bottom	5090 ± 80
HAR-1850	*Cervus* antler B 139/751, ditch (6/7)	4700 ± 80
CAR-820	*Cervus* antler 429, ditch (6/7)	4800 ± 80
CAR-819	*Cervus* antler 428, mound	4840 ± 70
CAR-818	*Quercus* + *Fraxinus* charcoal 350, ditch (5)	4450 ± 70
CAR-817	*Fraxinus* charcoal 237, ditch (5)	4370 ± 70
BM-2346	*Bos primigenius* tibia 355, ditch (4)	4120 ± 45
CAR-816	*Fraxinus* charcoal 232, ditch (4)	3830 ± 60

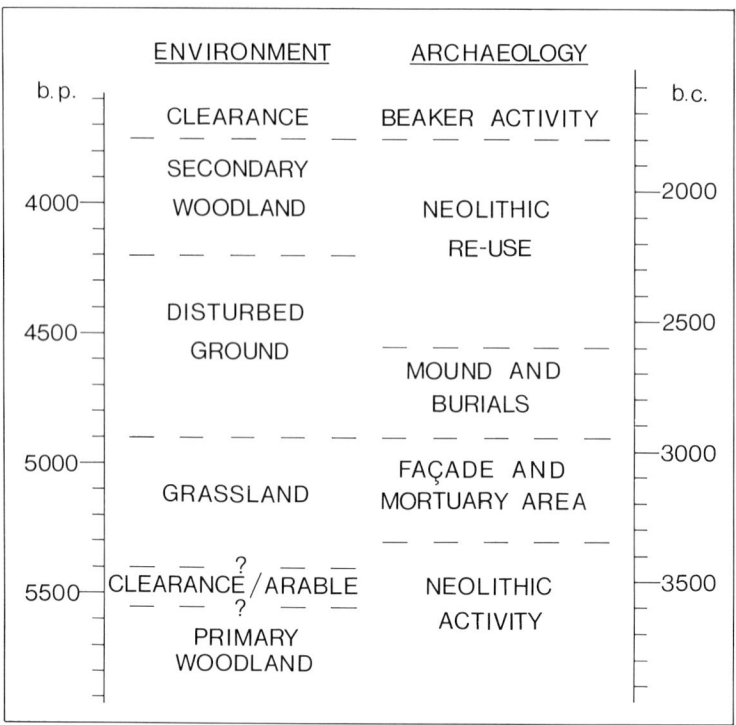

Fig. 5. Giants' Hills 2, Skendleby. Environmental and archaeological sequences for the Neolithic period. The question marks indicate that the clearance/arable episode is not directly related to radiocarbon dates.

proposed for other sites, for example for the Northumberland tomb of Street House, Loftus (Vyner 1984), but this is the first time that the chronology of the sequence has been securely dated.

The secondary re-use of the site is of considerable interest and may be compared to the activity represented by the infillings of chambered tombs, as for example in the case of the West Kennet Long Barrow (Piggott 1962). And the long time-scale of this secondary activity may be similarly compared to that proposed on archaeological grounds for West Kennet. The fact that re-use occurred largely in woodland is also noteworthy. There was a similar episode at the South Street Long Barrow in north Wiltshire, with later Neolithic occupation in a woodland context (Ashbee *et al.* 1979).

At South Street, too, it may be noted, there was clearance of secondary woodland in Beaker times, just as took place at Giants' Hills 2.

Many more radiocarbon dates for sequences where a combination of archaeology and environment may be set side by side in this way are urgently needed.

ACKNOWLEDGEMENTS

We thank the many individuals and institutions whose assistance has made this work possible. In particular we are grateful to Professor A.G. Smith and Quentin Dresser of the

University College Cardiff Radiocarbon Dating Laboratory, Dr J.A.J. Gowlett of the Oxford Radiocarbon Accelerator Unit, Richard Burleigh formerly of the British Museum Research Laboratory, and the C-14/Tritium Measurements Laboratory, AERE Harwell, for carrying out the radiocarbon measurements. The plant materials were kindly identified by Mr G. Morgan and Dr S. Limbrey. Quentin Dresser gave valuable advice during the writing of this paper.

REFERENCES

Ashbee, P., Smith, I.F. and Evans, J.G., 1979, Excavation of three long barrows near Avebury, Wiltshire, *Proc. Prehist. Soc.* 45, 207–300.

Barker, H., Burleigh, R. and Meeks, N., 1969, British Museum Natural Radiocarbon Measurements VI, *Radiocarbon* 11, 278–294.

Evans, J.G., 1972, *Land snails in archaeology*, London: Seminar Press.

Phillips, C.W., 1936, The excavation of the Giants' Hills long barrow, Skendleby, Lincolnshire, *Archaeologia* 85, 37–106.

Piggott, S., 1962, *The West Kennet Long Barrow: Excavations 1955–56*, London: H.M.S.O.

Vyner, B., 1984, The excavation of a Neolithic cairn at Street House, Loftus, Cleveland, *Proc. Prehist. Soc.* 50, 151–195.

RADIOCARBON DATING OF HAZLETON NORTH CHAMBERED TOMB: A PRELIMINARY STATEMENT

Alan Saville

The Neolithic Cotswold-Severn type chambered tomb at Hazleton near Cheltenham in Gloucestershire (NGR SP 073189) was totally excavated between 1979 and 1982 (Saville 1984). The tomb comprised a monumental long cairn containing a single pair of burial chambers, the latter located centrally within the cairn and entered through passages leading off the opposing long sides. Situated on either side of the long cairn were quarries from which the raw material was obtained for constructing the cairn (Fig. 1).

The excavation produced a large quantity of securely-contexted material suitable for radiocarbon assay. Most significantly this material included the extensive and well-preserved remains of numerous human burials in the chambered areas. As in most Cotswold-Severn tombs the skeletal remains of these collective burials were mainly in a disarticulated and scattered state but included were one complete and one partially complete inhumation in the north entrance.

The Hazleton human bones constitute an important assemblage of skeletal material, which offers the opportunity for multifarious scientific analyses in the long term (e.g. Rogers and Waldron 1985). The bones themselves are valuable archaeological objects and warrant careful conservation as part of the excavation archive. For this reason alone, conventional radiocarbon dating, which normally involves the destruction of entire bones, was inappropriate, whereas the Oxford accelerator, with its requirement of only tiny samples from individual bones, offered particular advantages. These include the ability to use the most substantial bones, such as longbones, which can be very anatomically informative (for example in isolating separate individuals), which are key bones in analyzing spatial distributions within the chambered areas, and which are large enough to permit the removal of several replicate accelerator samples without materially damaging the bone.

Other significant samples at Hazleton, for example from the buried soil, are of such a small size that only accelerator dating would be feasible if the now dubious practice of combining samples for conventional radiocarbon dating was to be avoided.

The initial programme of dating at Oxford involves about twenty samples from Hazleton, selected to investigate the internal chronology of the site. This chronology can be discussed in terms of the following four phases, designated (i) pre-cairn, (ii) monument construction, (iii) monument use, and (iv) monument decay.

(I) PRE-CAIRN PHASE

This phase is represented archaeologically by the artefacts and other remains preserved in the buried soil beneath the cairn. Some of the activity reflected by these remains could considerably pre-date the construction of the monument, especially in the case of the Mesolithic flint assemblage from beneath the west end of the cairn (Saville forthcoming). However, a midden area located centrally beneath the cairn just to the west of the burial chambers may have accumulated shortly before or even during the construction of the monument to judge from its position relative to the cairn (Saville 1984, Fig. 6) and from the artefacts it contained, including early Neolithic carinated pottery vessels. Animal bones were frequent in this midden deposit, but only in small fragments, two of which have been selected for dating.

The only date available so far for this pre-cairn phase comes from an isolated human cranial fragment found away from the midden beneath the west end of the cairn (Fig. 1). This date of 2925 ± 80 bc (OxA-646: 4875 ± 80) seems to relate to activity almost immediately preceding the construction of the cairn, in view of its near identicality with the date from the north chamber (see below). This raises the question of whether the cranium belongs to pre-cairn settlement activity (human skeletal fragments often being found at causewayed enclosures and other ostensibly non-funerary sites of this period), or to some mortuary activity previous to, or perhaps carried on during, the cairn construction.

Beneath the east end of the cairn the excavation of the buried soil recovered a single example of a grape-pip (*Vitis vinifera*). Such a find was unexpected for the early Neolithic (or an even earlier period) and since its context was unequivocal as far as the excavation records were concerned, it was clearly important to verify its actual date before considering the implications for prehistoric viticulture in England. The accelerator was able to date the pip and showed it to be modern (OxA-678), thus indicating that it must have been a contaminant, almost certainly finding its way on to the buried soil surface after exposure of that surface during excavation. What would otherwise have been an extremely enigmatic piece of data from Hazleton can thus be discounted thanks to the accelerator facility, which has a considerable potential for resolving such anomalous archaeological occurrences.

(II) MONUMENT CONSTRUCTION PHASE

The cairn itself contained virtually no organic material suitable for radiocarbon dating in contexts directly relevant to the monument construction. The only exceptions were two fragmentary antlers of red deer, which have not yet been included in the dating programme.

The primary fills of the two flanking quarries also produced numerous red deer antlers. These were presumably tools used by Neolithic workers in the quarrying process and can therefore be linked circumstantially to the construction phase. Many of these antlers are sufficiently large to be suitable for conventional radiocarbon dating, and it is intended to approach the dating of the Hazleton antlers in this way.

However, the implication of OxA-646 from the buried soil may be that a very high resolution indeed will be necessary to isolate a construction phase in between the pre-cairn and monument use phases, if the cairn was constructed rapidly over only a few years.

(III) MONUMENT USE PHASE

This phase is best documented by the burial deposits themselves, which were extensive and well-preserved with little evidence of later disturbance prior to excavation. Thirteen samples of human bones have been submitted for accelerator dating, deriving from

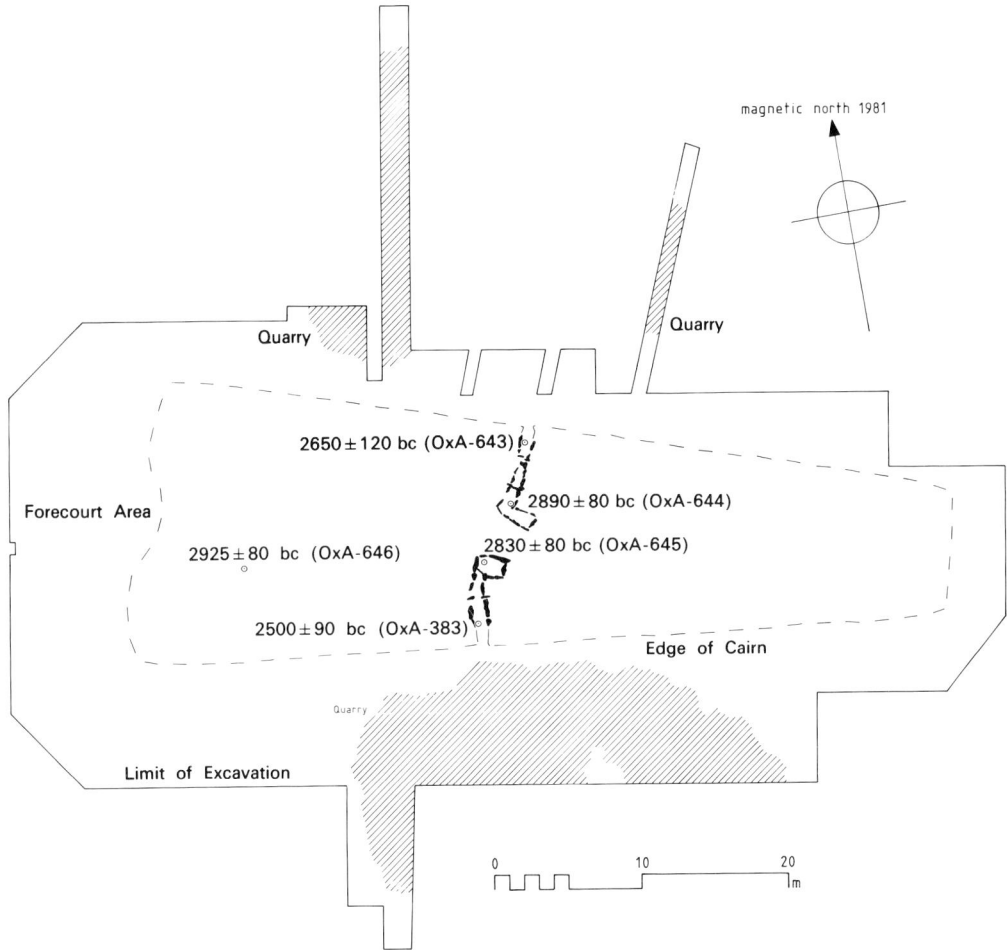

Fig. 1. Plan of the Hazleton North excavation area showing the extent of the cairn, the location of the chambered areas and quarries, and the radiocarbon dates available. Drawn by J. Hoyle.

different parts of the chambered areas as follows: north entrance 3; north chamber 2; south chamber 2; south passage 3; south entrance 3. The bones submitted are mostly femora, taken to represent different individuals, of which there are a minimum of 27 present on the basis of the number of separate skulls. However, one of the dates already obtained, from the south entrance, was of a human rib fragment since this was employed as an exploratory dating sample for Hazleton (Gillespie *et al.* 1985, p. 242). The dates so far available are listed below (see Fig. 1 for their locations).

OxA-644 femur, north chamber	2890 ± 80 bc	(4840 ± 80 BP)
OxA-645 femur, south chamber	2830 ± 80 bc	(4780 ± 80 BP)
OxA-643 femur, north entrance	2650 ± 120 bc	(4600 ± 120 BP)
OxA-383 rib, south entrance	2500 ± 90 bc	(4450 ± 90 BP)

The objective of the accelerator dating of burials at Hazleton is to investigate as far as is possible the overall duration of burial activity and the spatial organization of burial deposits within the burial areas through time. An independent control on the latter aspect was provided by the stratigraphic circumstances pertaining in the northern chambered area. The stratigraphy indicates that the northern chamber deposits had to pre-date those in the north entrance, and that at least two of the burials in the north entrance were clearly sequential (Saville 1984, pp. 17–19).

The accelerator dates so far obtained are very encouraging in suggesting a total date range which will be wide enough for patterning to be observable even at the two sigma level, and in demonstrating a chronological trend which accords with the stratigraphic evidence. Thus the results seem to suggest a period of several hundred radiocarbon years during which burials were made inside the Hazleton tomb, and that both north and south chambered areas were in use over the same period. The results also suggest that even on the south side, where through access was not restricted, the burials may be sequential from the chamber on into the entrance. However, caution is clearly required at this stage until confirmation of any of the above suggestions is available from further dates, especially as one of the existing dates has a regrettably wide standard deviation of ±120 radiocarbon years due to laboratory factors. (In this case the specimen will be redated, so that a higher precision result will be available from the combined values.)

Also probably associated with the phase of monument use is a deposit from the south quarry which included the fragmentary remains of a decorated 'Abingdon style' bowl and burnt and unburnt animal bone fragments. The vessel itself has a calcined bone temper, including fragments up to 10 mm in length, and wall sherds containing these bone fragments have been submitted for possible dating by the accelerator.

(IV) MONUMENT DECAY PHASE

This phase is inevitably lacking in contexts which are clear-cut and sealed and is therefore less susceptible to radiocarbon dating. The only contexts with suitable material of potential relevance were in the areas around the perimeter of the cairn, where the eventual collapse of the dry-stone revetment sealed artefacts on top of the Neolithic soil. This collapse is assumed to post-date the use of the monument for burial though this cannot be entirely vouchsafed, nor can the non-residual nature of the sealed artefacts be guaranteed. To investigate this problem two cattle skull fragments from beneath collapsed revetment in the forecourt area at the west end of the monument have been submitted for dating, and these will at least give a *terminus post quem* for the collapse.

In conclusion, it is hoped that the accelerator dating programme for Hazleton will make a real contribution towards resolving some of the outstanding problems associated with the construction and use of Cotswold-Severn tombs in particular, and British chambered tombs in general. With its extensive and fully recorded assemblage of human bones, Hazleton offers an appropriate test bed for the validity and potential of multi-sample high resolution dating to be applied to these still enigmatic monuments which are so crucial to our understanding of the earlier Neolithic period.

REFERENCES

Gillespie, R., Gowlett, J.A.J., Hall, E.T., Hedges, R.E.M. and Perry, C., 1985, Radiocarbon dates from the Oxford AMS system: Archaeometry datelist 2, *Archaeometry* 27, 2, 237–246.

Rogers, J. and Waldron, T., 1985, Lead concentrations in bones from a Neolithic long barrow, *J. Archaeol. Sci.* 12, 93–96.

Saville, A., 1984, Preliminary report on the excavation of a Cotswold-Severn tomb at Hazleton, Gloucestershire, *Antiq. J.* 64, 1, 10–24.

Saville, A., forthcoming, A Mesolithic flint assemblage from Hazleton, Gloucestershire, England, and its implications, *Proc. of the 3rd Int. Symp. on the Mesolithic in Europe, Edinburgh, 1985.*

RADIOCARBON AND THE CURSUS PROBLEM

Richard Bradley

Cursus monuments — otherwise cursuses or cursūs — are peculiar to British archaeology, and are among the most enigmatic features of the prehistoric landscape. They can be described as enormously elongated parallel-sided earthwork enclosures, from a few hundred metres to nearly 10 km in length, and never more than 100 m wide. Their chronology and their function still remain enigmatic, but their sheer scale makes them among the most elaborate constructions in the country. The Dorset Cursus, for example, represents a labour investment of more than 450,000 worker hours (Bill Startin, pers. comm.). There is no real consensus on the chronology of cursus moments, and those who do offer a view generally suppose that they belong to the Later Neolithic period and were associated with the celebration of the dead (e.g. Burgess 1980, pp. 338–339). Positive evidence has been very limited, however, consisting of a very few artefacts generally from ambiguous contexts in cursus ditches. These earthworks are occasionally associated with long barrows or oval crop mark enclosures of similar size to those monuments. In such cases the cursuses are usually thought to be a later development.

In fact their dating is not the only problem raised by cursus monuments. As earthworks whose construction must have made great demands on human labour, it is important to decide whether they are to be considered with the enormous henges dated to around 2000 bc, or whether they belong to an altogether earlier period of monument building, when causewayed enclosures and long barrows were being used. If the latter proved to be the case, it would significantly weaken the hypothesis that the scale of 'public' monuments increased steadily during the third millenium bc and that such a process provides one index of social evolution (Renfrew 1973). At the same time, the association of cursuses with long barrows, or with occasional discoveries of human bones, raises difficulties for the view that the Neolithic period in southern England saw a gradual change from collective to individual funerary monuments. This is particularly important now that the existence of Later Neolithic cremation cemeteries has been questioned (Bradley and Holgate 1984, pp. 123–126).

For both these reasons it is of paramount importance to establish the chronological context of these monuments. Artefact associations are not conclusive, but include material which could belong to the earlier third millenium bc, notably the excavated pottery from the Dorset Cursus (Bradley et al. 1984a, p. 90) and the Dorchester on Thames Cursus in the Upper Thames Valley (Atkinson, Piggott and Sandars 1951, p. 62). Apparently later material, including Beaker pottery, could suggest a long history for this form of monument. Alternatively, some sites could have been rebuilt or reused during later phases. At the same time, field survey seems to suggest a closer link with Earlier Neolithic funerary monuments than is often supposed. In particular, recent work in Dorset has emphasised the strong link between cursus monuments and bank barrows, and has also suggested that long barrows were still being built after both types of earthwork had been constructed (Bradley 1983).

The absolute dating evidence from these monuments is extremely limited, but it is entirely consistent. At present, dates are available for early contexts on the Dorset Cursus and two monuments in Oxfordshire, those at Drayton and Dorchester on Thames. A similar monument at Springfield in Essex has a *terminus ante quem* at the end of the third millenium bc (David Buckley, pers. comm.). In addition, a single date is available from the bank barrow at North Stoke on the Upper Thames gravels. In the near future this evidence should be supplemented by a sixth date from the Dorset Cursus and by two new series of dates – one submitted by the writer from early contexts associated with the cursus at Dorchester on Thames, and the other from the Lesser Cursus close to Stonehenge (Julian Richards, pers. comm.).

The largest group of dates comes from the Dorset Cursus, which is also the largest monument of this type. Excavation of the west ditch of this monument has shown that the primary silts had accumulated before the deposition of undecorated Neolithic pottery. Further excavation 700 m away showed that the secondary filling of the same ditch contained an assemblage of worked flints, animal bones, two human bones and sherds of Mortlake and Fengate Wares (Bradley *et al.* 1984b). At the time of writing five accelerator dates are available for samples collected during this work, and a conventional sample from the same series is being processed. The five samples so far dated come from the primary filling of the ditch, from a thin lens of humus on the surface of this layer and from the lower part of the secondary silts. Samples from the first two of these contexts consisted of small fragments of animal bone and returned dates of 4510 ± 140 bc and 4950 ± 100 bc respectively (OxA-628 and 627). Although no Mesolithic artefacts were found in the ditch excavation, an occupation site dating from this period is known a few hundred metres away and is perhaps the source of this material. An antler pick found on the surface of the primary silts is also to be dated (see note below). Two further pieces of animal bone came from the lower part of the secondary filling, below the level at which the pottery occurred, and accelerator dates for these give a combined estimate of 2625 ± 77 bc (OxA-625 and 626). The final sample from the middle of this layer gave a very similar date of 2620 ± 120 bc (OxA-624). Although one result is still pending, the three Neolithic dates agree in suggesting that the cursus may have been built in the earlier third millenium bc. In themselves these dates provide no more than a *terminus ante quem*, but molluscan analysis of the primary silts suggest that they had formed extremely rapidly.

The two samples from Drayton were also of animal bone, but this time they came from the primary filling of the cursus ditch (Richard Chambers and George Lambrick, pers. comm.). Since both dates refer to the same event, they can be combined to give a figure of 2935 ± 70 bc (HAR-6477 and 6478). No more samples are available from the initial use of this monument, but a useful comparison is with the single date from the primary levels of the North Stoke bank barrow, only 10 km away. A sample of antler from the bottom of the ditch here has a date of 2722 ± 49 bc (BM-1405; Case 1982).

The final date is also from the Thames gravels. This is an accelerator date of 2850 ± 130 bc for a human skull fragment found inside a pit close to the east end of the Dorchester on Thames Cursus (OxA-119; Bradley and Holgate 1984, p. 121). Until more samples from this monument have been processed, it is not certain where this feature comes in the sequence at this site, although it does seem possible that it was located close to the entrance of an early D shaped enclosure to which the main length of the cursus was added later (*ibid*).

There is no reason to suppose that these few dates reflect the full chronology of cursus monuments, although they do seem to demonstrate an earlier origin for such sites than has often been supposed. Additional samples will be dated in the near future, however, and these should provide more precise information. It is still possible that such monuments had a rather different history outside southern England, and this information should really be supplemented by dates from other areas. But taking the available evidence at face value, these dates do shed light on three issues mentioned earlier in this paper.

a)   They suggest that some cursuses, including the largest of all these earthworks (the Dorset Cursus), were built during the same phase of monument construction as long barrows and causewayed enclosures.

b)   They suggest that there may have been a more direct link between some of these different classes of monument than is commonly supposed. In particular they emphasise the apparent connection with long barrows, whose own period of construction may have ended about 2500 bc.

c)   Lastly, they suggest that cursus monuments may initially have been with the large scale celebration of the dead, so characteristic of Earlier Neolithic society in Britain. Despite the proximity of round barrows, ring ditches and hengiform enclosures to some of these sites, they are not directly connected with the development of individual burial found in the Later Neolithic period.

These dates represent only the first steps in resolving the cursus problem and further work must take a more sophisticated attitude to current debates in theoretical archaeology. But by establishing the chronological context of these sites they make it possible, as never before, to speculate about the social forces that lay behind the building of such outlandish monuments.

NOTE: The antler pick mentioned above has now given a date of 2540 ± 60 bc (BM 2438). This agrees well with OxA 624–626.

REFERENCES

Atkinson, R.J.C., Piggott, C.M. and Sandars, N.K., 1951, *Excavations at Dorchester, Oxon*, Oxford: Ashmolean Museum.

Bradley, R., 1983, The bank barrows and related monuments of Dorset of the light of recent fieldwork, *Proc. of the Dorset Nat. Hist. and Archaeol. Soc.* 105, 15–20.

Bradley, R., Cleal, R., Gardiner, J., Green, M. and Bowden, M., 1984a, The Neolithic sequence in Cranborne Chase, in *Neolithic Studies* (eds. R. Bradley and J. Gardiner), pp. 87–105, Oxford: B.A.R. 133.

Bradley, R., Cleal, R., Gardiner, J., Legge, A., Raymond, F. and Thomas, J., 1984b, Sample excavation on the Dorset Cursus, 1984 — preliminary report, *Proc. of the Dorset Nat. Hist. and Archaeol. Soc.*, 106.

Bradley R, and Holgate, R., 1984. The Neolithic sequence in the Upper Thames Valley, in *Neolithic Studies* (eds. R. Bradley and J. Gardiner), pp. 107–134, Oxford: B.A.R. 133.

Burgess, C., 1980, *The Age of Stonehenge*, London: Dent.

Case, H., 1982, The linear ditches and southern enclosure, North Stoke, in *Settlement patterns in the Oxford region* (eds. H. Case and A. Whittle), pp. 60–75, London: Council for British Archaeology.

Renfrew, C., 1973, Monuments, mobilisation and social organisation in Neolithic Wessex, in *The explanation of culture change*, (ed. C. Renfrew), pp. 539–558, London: Duckworth.

RADIOCARBON: A MEANS TO UNDERSTANDING THE ROLE
OF BRONZE AGE METALWORK

S. P. Needham

Of all the prehistoric material from north-west Europe little before the Late Iron Age has been given such refined dating as Bronze Age metalwork.

From the first formulation of a prehistoric chronology until the advent of radiocarbon dating, Europe depended on a chronology transposed from historical cultures in the East Mediterranean and the Near East. In time this led to the development of complex schemes, central to which in the Bronze period were grave assemblages, which often contained diagnostic metalwork. Where the burial record gave out, as in much of Bronze Age Britain, the most detailed and reliable chronology was hung on the metalwork itself, both hoards and diagnostic single finds, by means of a series of inter-regional stylistic links and exchanged objects (e.g. Butler 1963; Burgess 1968). The chronology devised was absolute and rigidly conceived; it has for the most part remained unchallenged even after thirty years of radical revision at the hands of conventional C14 dating. For example, the substantial back-dating of some Early Iron Age settlement evidence into the full Late Bronze Age (e.g. Barrett 1980a), or the realisation that the Deverel-Rimbury culture began during the Early Bronze Age rather than strictly succeeding it (Barrett 1980b, pp. 82–83, fig. 2), have not in themselves altered the underlying framework of the traditional chronology. In effect this has led to the polarisation of two independent chronologies, the traditional and the newly constructed C14 schemes, which are applicable respectively to metalwork and settlement evidence. Any attempt at marrying the two has in the past been hamstrung by the infrequency of metalwork in domestic contexts and now continues to be hampered by the lack of association between metalwork and C14-dated material. Rather few of the many conventional C14 dates for the period can be applied to Bronze Age metalwork, and fewer still are in *direct* association (Fig. 2).

Here then is a potential block to the full integration of Bronze Age cultural systems which might be soluble by a programme of C14 analysis directed expressly towards the dating of bronzes. However, before advocating such a course it is worthwhile examining carefully the current issues of concern and the prospect of resolving them.

The chronological framework for Britain has evolved into an elaborate scheme (Fig. 1; Burgess 1979, p. 210, fig. 2; Burgess 1980, p. 270, fig. 15) which does more than justice to Akerman's observation in 1850 that "the period termed by antiquarians the Bronze Age is susceptible of more than one division" (*Proc. Soc. Antiq. London* 2, 1849–53, p. 168). It would probably be generally agreed amongst bronze specialists that the sequence is correct in broad outline and that there are limited prospects for sub-division. So what do we actually want to learn about this material and what do we believe the existing framework to represent? The structure could be accepted as portrayed here (Fig. 1), a straightforward chronological development of the metalwork, allowing just a little overlap between 'stages';

Date BC		Stages (after Burgess)	Metalworking Assemblages (*after Needham*)
?	Copper	I Castletown Roche	MA I
		II Knocknague	
			MA II
		III Frankford	
	EBA	IV Migdale-Killaha	MA III
2000			
		V Aylesford-Colleonard	MA IV
1800			
		VI Bush Barrow-Willerby Wold	MA V
1650			
		VII Arreton-Inch Island	MA VI
1500			
	MBA 1	VIII Acton Park	
1350			
	2	IX Taunton	
1200			
	3	X Penard	
1000			
	LBA 1	XI Wilburton-Wallington	
900			
	2	XII Ewart Park	
700			
	3	XIII Llyn Fawr	
650			

Fig. 1   The British sequence of metalwork traditions. The dates given are in calendar years and represent recent views of the chronology assessed by traditional means. Suggestions of phase overlap within the schemes presented have been omitted for the sake of clarity.

different 'industries' come and go, are responsible for the introduction of new technology and types, and have changing regional foci.

However there is increasing opposition to this traditional interpretation and increasing debate about the nature of production, circulation and particularly deposition (e.g. Bradley 1982; Needham forthcoming). One acute problem in interpretation, made clear by recent discoveries and research, is that the bronzes accumulated through casual finds over the past two centuries are far from representative of the full Bronze Age range; a given type of hoard may be an extremely partial record. Current controversies do not deny a broad temporal succession from top to bottom of the chart, but they question relationships of all kinds between adjacent (or even separated) boxes in the chart, especially since the defining categories are not always based on like material and contextual evidence. For example, there has been a long-lived debate on the chronological relationship between Wessex 1 and

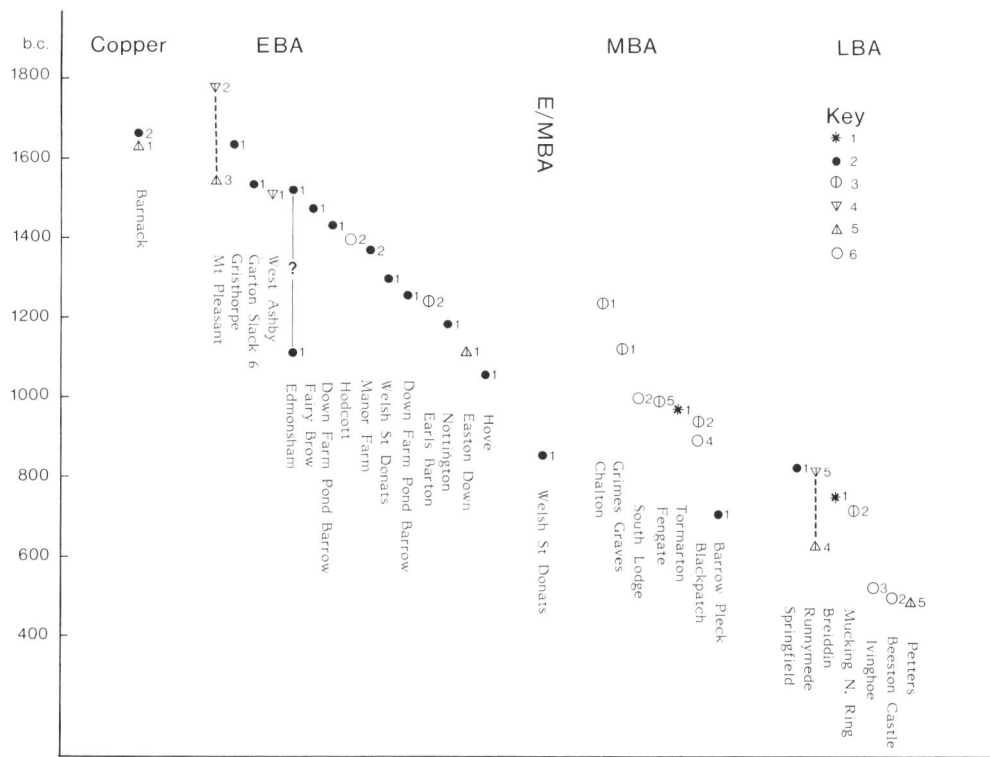

Fig. 2 Conventional C14 dates associated with Bronze Age metalwork. The numbers accompanying symbols refer to the number of determinations combined as a single mean. Key to symbols: type 1: C14 sample(s) part of the artifact (e.g. haft) 2: Sample(s) and artifact in closed context (e.g. grave) 3: Sample(s) and artifact in similar stratification 4: Sample date gives TPQ for artifact 5: Sample date gives TAQ for artifact 6: Sample(s) and artifact in loose association on a site

2 graves (EBA VI & VII) which are characterised *inter alia* by different metalwork forms, but also could be seen to represent distinct social strata. Another question might be how the weapon hoards of Penard type, nominally MBA3, relate to the ornament and tool dominant hoards of the Taunton 'stage', MBA2. Again Wilburton material (LBA1) tends to be stongly clustered in certain regions and could thus be seen to be a specialised, resticted and short-lived industry within an emerging Ewart Park tradition (LBA2). This last example highlights above all the difficulties of interpretation in a contextual and temporal vacuum and the need for independent dating. It poses the question, to what extent is the regionalisation observed either a true reflection of zones of output and circulation, or by contrast a product of dichotomies between metal-depositing and metal-conserving societies, themselves perhaps due to current differential access to material surplus? At present these complex four-way interactions between time, location, artifact type(s) and

context can usually be distorted at will to suit the individual argument (c.f. Kristiansen 1985, pp. 251–252). Radiocarbon may at least enable one variable to be more positively fixed. The artifactual entities, which the writer prefers to call *Metalwork Assemblages*, a term without prejudice to chronology and industrial background (Needham *et al.* 1985), are nevertheless *bound* to have a chronological and spatial definition, but for the present the limits should be treated as fluid.

On a practical level, there are virtually no accelerator dates that relate to bronzes, but by considering relevant C14 dates problem areas can be isolated and illustrated. Potential sources of error are manifold. A prevalent problem for existing data is imprecise association between dated samples and their respective artifacts, and another is the undiagnostic nature of many of the bronzes in question. In addition to these are further contextual problems – the possibility of residual artifacts (particularly on settlements) or residual charcoal – and of course all the C14 specific problems, from standard deviation to calibration.

In Fig. 2 multiple series of dates from individual sites have been combined as one or more appropriate mean values, while dates with standard deviations of above ± 100 years have been excluded, unless supported by a second compatible date on the same context. The inadequacy of the available data is well emphasised. For the Late Bronze Age one multiple date merely serves as a *terminus ante quem*, whilst another is very loosely associated with the bronzes on the site. Again few good associations exist in the Middle Bronze Age group, and single determinations at top and bottom could easily be regarded as outliers. The Early Bronze Age gains from having a higher percentage well associated, but loses out in that most sites have single dates. One of few sets of multiple dates is the pair from the Edmonsham grave which should date contemporary contexts. However, as the two dates differ by as much as 400 years, one of very few dated Wessex culture graves has to be disregarded.

We have seen then that the major period divisions are held in reasonable sequence by associated C14 dates, but ideally it is necessary to look at the finer divisions, the Metalwork Assemblages, given the sort of questions raised above. Sadly most of the metalwork in the lengthy EBA bracket is undiagnostic (Fig. 3). The examples could be construed as presenting a coherent sequence of the sort anticipated, with the exception of a new and disappointing result from Manor Farm. The MBA results also look reasonable, given that all are loose associations from settlement contexts. They might be regarded as having a major overlap with EBA4. This view, however, would be questionable on other grounds (see below) even though various recent research has pointed to overlap between final EBA burial traditions and early MBA material culture. Dates for the Ewart Park phase (LBA2), which is thought to occupy most of our Late Bronze Age, sit happily between 850 and 600 bc, with a possible later continuation. As with MBA1, however, there are serious gaps for LBA1 and LBA3, thus precluding comparisons.

At this stage it is worth investigating the limitations on achieving the required resolution. One way to do this is to perform in effect a reverse calibration which allows the definition of *expected* C14 ranges based on the currently accepted calendrical chronology for bronzes (i.e. that dervied from historical chronologies) and the newly established Belfast calibration curve (Fig. 4, Pearson *et al.* forthcoming). To take the major period divisions, it is assumed for this purpose that there was no overlap, as depicted along the top of Fig. 4. In converting their limits to C14 dates the ranges have to be broadened to accommodate the reasonable

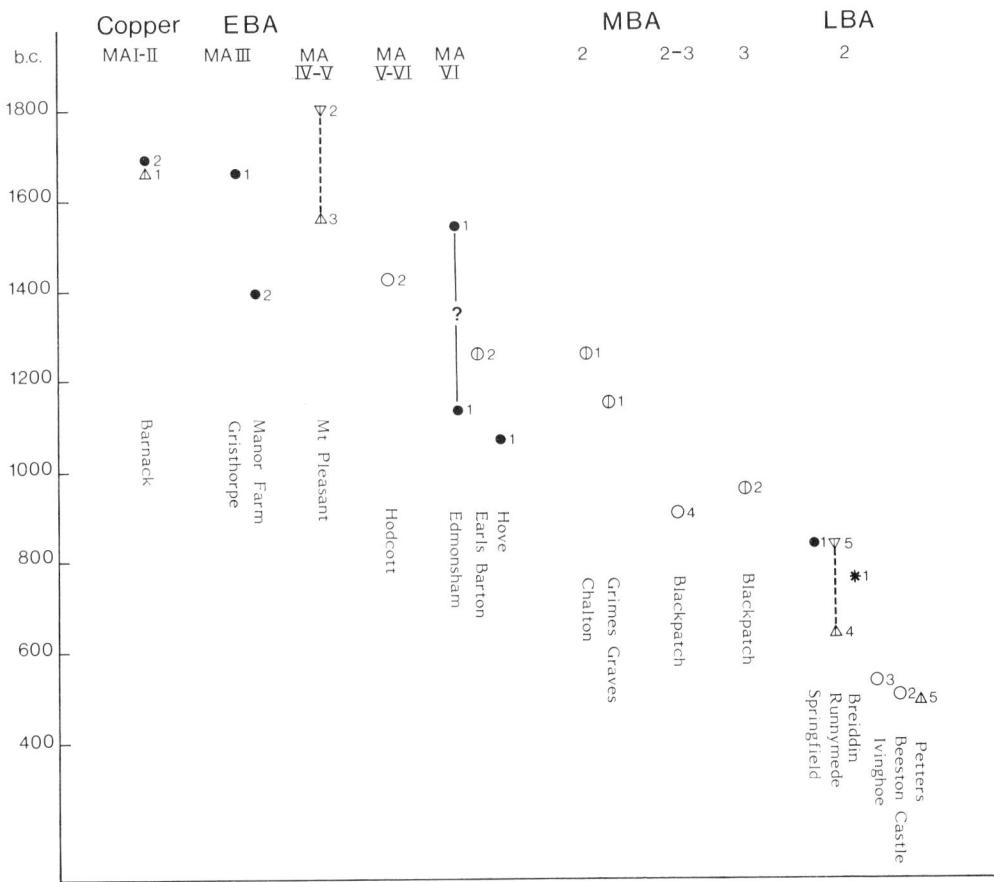

Fig. 3 Conventional C14 dates associated with Bronze Age metalwork grouped according to 'stages' or Metalwork Assemblages. Key as for Fig. 2.

possibility of some very early or very late determinations falling outside the expected bracket by as much as one standard deviation. The proportion of determinations lying still further outside may be taken to be negligible. An arbitary standard deviation of ± 50 years has been used to cover the tail ends of the distributions for these periods. It is noteworthy that in this predictive model the EBA and LBA ranges are separated by only about one hundred C14 years. The actual conventional dates plotted in Fig. 2 almost entirely fall within these predicted ranges. Consequently, given the current limited and problematical sample, it is not yet possible to claim any overlap between the major period divisions.

The effect of doing a similar reverse calibration on the finer divisions, which are of more interest, would be firstly and obviously to reduce the individual expected C14 ranges according to accepted phase duration and the nature of the calibration curve for the period, but secondly to change the shape of the expected distributions which would theoretically move from plateau-like forms towards more peaked forms. A useful indication of the tightness of clustering to be expected for a Metalwork Asssemblage of limited duration can be gained from Orton's work (1983). He states that "for relatively short phases (1–2 standard deviations in length), about half of the observed dates can be expected to lie outside the true range of their phase" (Orton 1983, p. 119). This would apply, for example, to a assemblage believed to have a currency of about one century, although it does not take account of the complexity and variability of the calibration curve. It does emphasise, however, that in order to achieve the resolution desired in metalwork studies the number of dates required for each assemblage of interest should be statistically viable, since interpretation is likely normally to hang on comparison of distribution shapes as well as their ranges. Consequently choice of samples becomes critical and this is why the accelerator technique offers so much better prospects than conventional C14 dating, for most of the archaeological sources of uncertainty can be systematically eliminated by the capability of dating material in direct association with the bronze artifacts. With the judicious selection of bronzes having attached organic remains, whether hilt, haft, shaft, thong, sheath or textile wrap, the likelihood of any significant time lapse between organic sample and metal component is minimised. Furthermore, the small sample size required for accelerator dating should make this a viable proposition even for well preserved organics which are important technological documents in their own right.

Of the material available, pre-eminent in the Early Bronze Age are the sheath remains often encountered on grave-found daggers, which are generally diagnostic types. Using such material it would, for example, be possible to rectify the complete lack of dates for the Armorico-British series of Wessex 1 (*pace* Bradley 1984, p. 89) and accumulate a decent set for the nominally earlier flat daggers and the later Wessex 2 examples. For the Middle and Late Bronze Age emphasis would have to shift to the many river and bog deposited bronzes, notably those with sockets, which not infrequently contain shaft/haft tips. Again such a programme should concentrate on the more diagnostic types.

If the initial aim of providing a more secure chronology for Bronze Age metalwork seems pedestrian, its fulfilment would nevertheless automatically open other doors. In particular we would stand to gain a deeper understanding of the social and functional reasons behind deposition based on a temporal framework which was not interdependent with interpretation itself. This would constitute a major contribution to knowledge of Bronze Age Society.

ACKNOWLEDGEMENTS

The raw data of this paper inevitably includes as yet unpublished radiocarbon dates; my thanks go to those who provided details of them. As for the interpretative side, valuable advice was offered by Morven Leese on statistical matters and Ian Longworth on archaeological presentation. Philip Compton kindly assisted with the preparation of diagrams.

REFERENCES

Barrett, J., 1980a, The pottery of the Later Bronze Age in lowland England, *Proc. Prehist. Soc.* 46, 297–319.

Barrett, J., 1980b, The evolution of Later Bronze Age settlement, in *Settlement and society in the British Later Bronze Age* (eds. J. Barrett and R. Bradley), pp. 77–100, Oxford : *BAR British Series* 83, 2 vols.

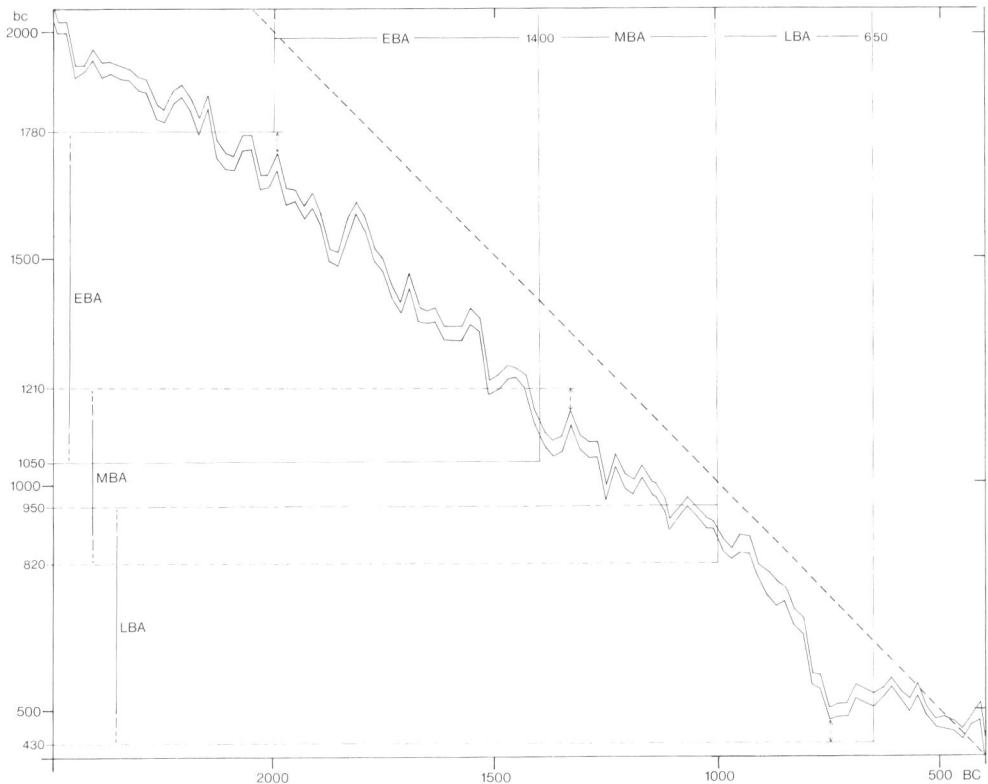

Fig. 4 The calculation of expected C14 ranges (y-axis) for the main period divisions using the Belfast calibration curve (Pearson et al., forthcoming). A standard deviation of ± 50 has been added to the beginning and end of each period.

Bradley, R., 1982, The destruction of wealth in later prehistory, *Man* 17, 108–122.

Bradley, R., 1984, *The social foundations of prehistoric Britain: themes and variations in the archaeology of power*, London: Longmans.

Burgess, C.B., 1968, The later Bronze Age in the British Isles and north-western France, *Archaeol. J.* 125, 1–45.

Burgess, C.B., 1979, The background of early metalworking in Ireland and Britain, in *The origins of metallurgy in Atlantic Europe* (ed. M. Ryan), pp. 207–214, *Proc. of the 5th Atlantic Colloquium, 1979, Dublin*.

Burgess, C.B., 1980, A find from Boyton, Suffolk, and the end of the Bronze Age in Britain and Ireland, in *Bronze Age hoards: some finds old and new* (eds. C.B. Burgess and D. Coombs), pp. 269–283, Oxford: *B.A.R. British Series* 67.

Butler, J.J., 1963, Bronze Age connections across the North Sea, *Palaeohistoria* 9.

Kristiansen, K., 1985, The place of chronological studies in archaeology: a view from the Old World, *Oxford J. Archaeol.* 4, 251–266.

Needham, S.P., forthcoming, Selective deposition in the British Early Bronze Age, in *Proc. of the British-Scandinavian Bronze Age Colloquium, Stockholm, May 1985 (ed. H.A. Nordström)*, *Stockholm Studies Series*.

Needham, S.P., Lawson, A.J. and Green, H.S., 1985, *Early Bronze Age hoards*, London: *British Bronze Age Metalwork: Associated Finds Series* A1–6.

Orton, C., 1983, A statistical technique for integrating C14 dates with other forms of dating evidence, in *Proc. of the Conf. on Computer Applications in Archaeology* (ed. J. Haigh), pp. 115–124.

Pearson, G.W., Pilcher, J.R., Baillie, M.G. and Corbett, D.M., forthcoming, High precision ^{14}C measurement of Irish oaks to show natural ^{14}C variations from 5000 BC to 1840 AD, in *Proc. of the 12th Int. Radiocarbon Conf., Trondheim, Norway, June 24–28, 1985* (eds. M. Stuiver and R. Kra), *Radiocarbon* 28, 1.

METHODOLOGICAL ISSUES IN THE STUDY
OF BRONZE AGE CHRONOLOGY

Richard Bradley

'Hands, do what you're bid:
Bring the baloon of the mind
That bellies and drags in the wind
Into its narrow shed.'

W.B. Yeats

Until relatively recently the study of Bronze Age chronology *was* the study of the Bronze Age. Questions of dating dominated any other type of activity and discussion of more ambitious problems was deferred until the ideal ordering of the evidence had been achieved. Since that objective has still to be attained, it may be well to begin this piece by asking why we should be interested in the Bronze Age at all.

Apart from questions of chronology, this period allows us to investigate at least five themes of general interest. Although all of these projects necessitate improvements in chronology, none will be achieved by finer dating alone.

1. Bronzeworking is a relatively complex technology, compared with the use of stone and to some extent with the use of iron. The restricted distribution of the raw materials, and the attractive appearance of the finished products, both make bronzes an ideal medium for display. As a result, the process of social change is especially conspicuous in the archaeology of this period.

2. The need to distribute both raw materials and finished products means that long distance exchange also plays a major part during this phase. This introduces several themes of much wider significance, in particular the relationship between the development of exchange and the growth of social complexity; the distinctive character of high status or other specialised spheres of exchange; and in certain areas the relationship between early complex societies and the barbarian periphery which may have supplied them with some of their metals.

3. Bronze artefacts may be recycled, so that a large number of those which now survive must have been deposited deliberately — in burials, hoards, or so-called votive deposits. This means that the Bronze Age affords an opportunity to investigate the consumption of wealth in non-market societies. Indeed, the sacrifice of so many objects could shed light on the concepts of wealth and value in such communities.

4. A significant proportion of complex artefacts occur in Bronze Age burials, one of the most significant arenas for the display of valuables. This period allows us to consider the relationship between grave goods and social organisation, particularly since Shennan has argued that the adoption of metallurgy in Western Europe was accompanied by an ideology which encouraged the display of individual wealth and power (1982).

5. The Later Bronze Age saw a drastic reorganisation of agricultural production. In view of the comments made already, this period provides one of the best opportunities for studying the relationship between changes in food production and ideological and industrial changes reflected by the production and consumption of metalwork.

Although the Three Age System itself is out of date, these five approaches do suggest that the Bronze Age still has something to offer prehistoric archaeology. Having criticised the overemphasis on chronological studies in this period, it may seem rather perverse to add the rider that the more ambitious programme suggested here still requires a detailed appreciation of the Bronze Age sequence. But such an appreciation must be independent of the very processes which I have already described. This has not been the case, and it is why the literature on this period possesses such a hermetic character. For this reason it is very difficult to bring Bronze Age studies into line with other current research.

If Bronze Age scholars are to contribute to a more broadly based archaeology they must take a step backwards before they can move forward, for the chronological framework employed at present betrays their innocence of the very problems which make the period so interesting. This is why objective analysis requires the precision of radiocarbon dating.

There are four main difficulties in compiling a chronology by the methods used at present. To a large extent these result from the fact that most schemes are based on metal types which do not occur in large numbers on settlements, with the result that it is necessary to integrate distinct sequences which are founded on very different assumptions. Settlement evidence is generally fitted to metalwork chronologies based on the evidence of burials and hoards, but there is little to indicate a straightforward relationship betwen stylistic change and the passage of time. Work in Scandinavia has shown that bronze artefacts could have circulated for quite different periods before their deposition (Kristiansen 1978), and in Britain the same types of artefact in Bronze Age hoards can show different amounts of wear from one region to another (Robin Taylor, pers. comm.). The same *types* of object could also change their roles. This argument has already been applied to Early Bronze Age pottery (Bradley 1984, pp. 70–73), but it also applies to metalwork. For example, the earliest daggers in southern England may have been treated as insignia and could have had a large symbolic function, with the result that they show less signs of wear than those deposited during the Wessex Culture (Julia Wall, pers. comm.). Other metal items may have been made for use in funerals and could have been buried at once.

At a more general level, there are dangers in erecting a metal chronology on the evidence of hoards, when we cannot agree what hoards mean, why they were deposited in the first place or why their contents were never recovered. There is a possibility of chronological development here, with the use of essentially votive deposits in the Early Bronze Age and the storage of commodities in the Late Bronze Age. Still more troubling is the suggestion that hoarding took place during periods of stress and that such deposits included types of artefact which were going out of style (Bradley 1985, p. 39). Indeed, these hoards could have survived entirely by accident and their contents may initially have been stored for later recovery. If so, we have built a complete chronology from material whose survival was peculiar to periods when the cycle of storage and retrieval was disrupted. In any event, the main characteristic of the 'votive hoards' in parts of Europe is the specialised nature of the metalwork, which means that it may not have been used in the domestic sphere.

Further problems arise because bronze artefacts are so suitable for displaying status. Many of these finds come from graves, but there seems no reason to suppose that the lavish consumption of metalwork during funerals was a regular feature of the Bronze Age. Rather, the provision of lavish burials, like the building of elaborate monuments to the dead, may be a discontinuous phenomenon, found at times of competition or rapid change (Parker Pearson 1982). For example, the chronology of Early Bronze Age Britain depends on finds of pottery and metalwork from graves, yet the available radiocarbon dates suggest that the more elaborate burials belong to two main periods separated by a significant hiatus (Bradley 1984, p. 89).

Lastly, differential access to elaborate metalwork could mean that change took place at different rates in different areas of Europe. Not every community had access to the latest developments, and social pressures could have been exerted to maintain their exclusive character. Such contrasts cannot be revealed by chronological schemes which depend on one all-encompassing arrangement of the artefacts. For example, it has only been the application of radiocarbon dating to settlements and cemeteries in Later Bronze Age Britain that has shown how far developments in Wessex lagged behind those in the Thames Valley (Bradley 1984, pp. 106–124). Application of a more rigid chronology, in which typological changes moved at a uniform pace, would have displaced some of the settlements by several hundred years.

For all these reasons a radiocarbon chronology is a prerequisite if we are to undertake the more abstract studies for which the Bronze Age is so well suited. But we must work on a uniform basis and the sequence of metal types, which still retains its primacy, must be tested by radiocarbon; for only in that way can we compare the evidence from excavated settlements with the traces of conspicuous consumption which have provided so much of our raw material. This is possible using accelerator techniques, since these can sample small parts of valuable artefacts, or the surviving traces of their hafts, and can then date them on the *same scale* as the settlement sites or graves. In taking this approach we can free ourselves of the uniformitarian assumptions which underpin European chronology.

Such a modest proposal may seem a damp squib after so much polemic, but the same programme of work could lead us back to some of the most complex questions in prehistoric archaeology. An accurate and comprehensive chronology for the richer burials, compiled from the artefacts themselves and also from samples of human bone, would do more than place them in the correct alignment with their European counterparts. It would shed light on conspicuous burial as a phenomenon worthy of study in its own right, and could help us to discover the relationship between the display of social position and contemporary developments in the pattern of settlement. In exactly the same way, careful study of the elaborate objects deposited in rivers and bogs could do more than track different classes of exotica around Bronze Age Europe. This work could shed light on the pace and intensity of consumption during what is by any account a time of drastic change. This could be effected by radiocarbon dating not only of the hafts of those artefacts, but also of the bones which were sometimes deposited with them. Again, reliance on a uniform value-free method of dating could allow us to consider the role of conspicuous consumption alongside the evidence of more open competition afforded by the earliest hill forts.

So if radiocarbon dating can bring order where now there is confusion, it can also become an analytical tool of some sophistication. In cutting out the most laborious stage in Bronze Age research, it could also rejuvenate both the study and the students of this period.

R. Bradley

REFERENCES

Bradley, R., 1984, *The Social Foundations of Prehistoric Britain*, Harlow: Longman.

Bradley, R., 1985, The archaeology of deliberate deposits, in R. Bradley, *Consumption, change and the archaeological record*, pp. 21–40, Edinburgh: Edinburgh University Department of Archaeology Occasional Paper 13.

Kristiansen, K., 1978, The consumption of wealth in Bronze Age Denmark, in *New directions in Scandinavian archaeology, Vol. 1* (eds. K. Kristiansen and C. Paludan-Müller), pp. 158–190, Copenhagen: National Museum of Denmark.

Parker Pearson, M., 1982, Mortuary practices, society and ideology: an enthnoarchaeological study, in *Symbolic and structural archaeology* (ed. I. Hodder), pp. 99–113, Cambridge: University Press.

Shennan, S., 1982, Ideology, change and the European Bronze Age, in *Symbolic and structural archaeology* (ed. I. Hodder), pp. 155–161, Cambridge: University Press.

Section V

Radiocarbon Methodology and Technology

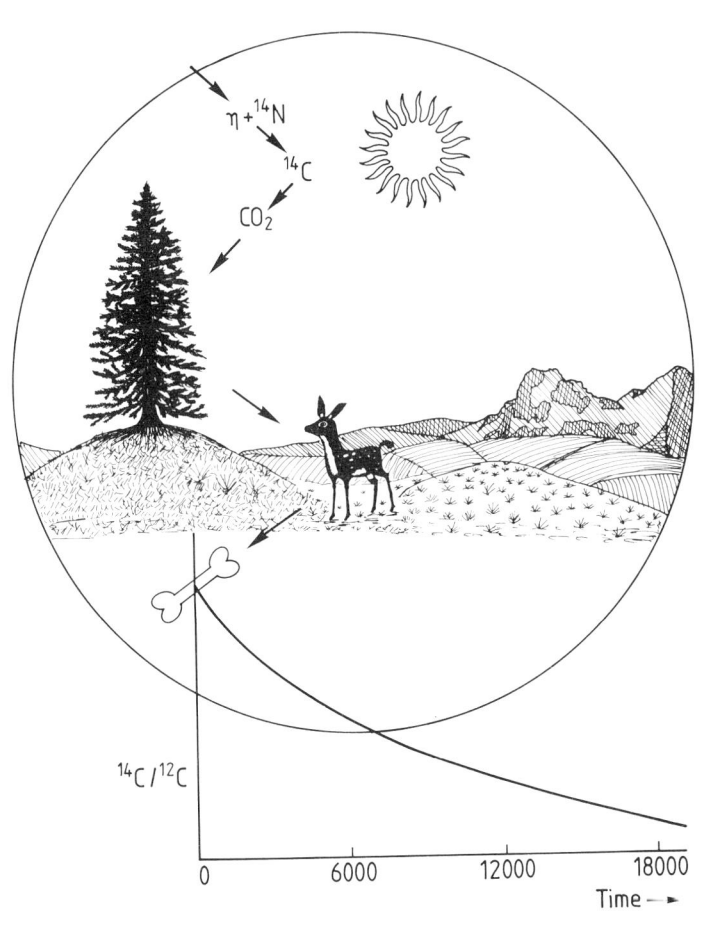

THE TERMINOLOGY OF TIME

R. Gillespie and J. A. J. Gowlett

INTRODUCTION

In the last thirty years, radiometric dating methods have entirely altered the scope of archaeological and geological interpretation. Yet this transformation has been achieved only with great effort and constant debate, and in large measure its course has been unpredictable. Where the course of developments is uncertain, terminology has a habit of becoming outdated and misleading, and if that is so for the harder sciences, such as chemistry, it is not surprising that we have become beset with difficulties in the area of dating for archaeology.

From 1946, when the first edition of Zeuner's *Dating the Past* appeared, the long term prospect was already plain: the long epochs of geology could be dated by radiometric means, and given time and luck, methods would be developed to cope with the time-range of prehistory. But naturally enough, the majority of archaeologists were not fully prepared for this to happen in practice, nor were physicists always able to anticipate every limitation in their techniques, some of which could cause archaeological confidence to falter. In general that confidence has been won back by consistent patterns of results, and fuller explanations, but the very interests of consistency in expressing results have made it difficult to make alterations to conventions, or to experiment with finding the best approach to satisfying workers in several disciplines.

At present there are several important physical methods, with different constituencies of archaeologists, and sometimes with potentially conflicting descriptions of their timescales. In this paper we aim to sketch out the historical course of developments, the nature of the unforeseen problems and their consequences, together with some possible directions for future consensus.

SCIENTIFIC TERMINOLOGY: INTERNATIONAL UNITS AND THE ESTABLISHMENT OF STANDARDS

The international unit of time is the second (s). The year is not an absolute unit, but is sufficiently close to it to be allowed in the SI system of units, and must be our mainstay. The symbol for it is a (annum). It can therefore be combined with other symbols designating larger values, principally:

ka — thousands of years

Ma — millions of years

Other commonly used symbols such as kya and Myr are less correct, and their use should be discouraged. The only other widely accepted reference points for time are those designating historical eras: of these the Christian calendar of AD/BC is so well known as to need no discussion in itself. But it is worth noting that the irrelevance of this schema in areas where prehistory extends almost up to the present has encouraged the usage of a 'Before Present' notation (discussed below at greater length).

THE EARLY APPLICATION OF RADIOCARBON

Radiocarbon is the most familiar 'absolute' dating method, and has therefore had a large influence on dating terminology. Ironically, some of its limitations only became apparent as the database of results grew, and many of the terminological problems which we have are historical ones resulting from the gradual appreciation that the primary absolute method is not truly absolute. There are now over 65,000 radiocarbon dates, of which at least 30,000 are relevant to archaeology (R. Burleigh, pers. comm.).

Definitions in radiocarbon usage have emerged in a fairly haphazard way over the years, but in a broad sense they have been concordant with the principles of scientific nomenclature, and have been accepted and ratified at International Conferences of radiocarbon daters — the only effective forum for such discussions. Most of the important decisions have been published subsequently in *Radiocarbon*, which first appeared in 1959 (see, e.g. Flint and Deevey, 1961 1962 and 1963).

Radiocarbon dates are calculated from the ratio of ^{14}C specific activity of a material of interest with respect to a modern reference standard. Wood seemed an ideal material for this, but it soon became apparent that the industrial carbon dioxide in the atmosphere (the dead fuel effect) ruled out trees grown this century.

Thus radiocarbon dates were originally calculated using 'contemporary' wood (variously 50, 80 or 100 years old) as the modern standard. Since the decay of this standard up to the year of dating was taken into account, a date produced in 1958 was effectively 'Before 1958'. But after the Groningen conference of 1959, NBS oxalic acid was adopted as the standard, with the advantage that it was homogeneous and could be distributed to the various laboratories. Soon afterwards a defined zero year of AD 1950 was adopted for the expression of all dates (Flint and Deevey 1962).

A second fundamental constant in radiocarbon dating is that of the half-life. Initially, the original Libby value of 5568 years for the half-life of ^{14}C was used (Libby 1952, 1955). Later a more accurate value of 5730 years became available (Mann *et al.* 1961) but it was decided at the Pullman conference of 1965 to retain the original Libby value so as to avoid confusion in the citation of dates. Use of the longer half-life would increase radiocarbon dates by 30 years in 1000, but it soon became apparent that other factors were distorting radiocarbon years from historical years in a way which made this almost irrelevant.

It had been implicitly assumed that radiocarbon ages were measurements on an absolute scale of calendar years, but measurements on known-age materials such as tree-rings soon showed that this was in fact incorrect (De Vries 1958) and the need for calibration scales became apparent. Prior to this, radiocarbon ages had been converted to the AD/BC scale by simply subtracting 1950 years from the calculated date, though the absurdity of '40,000 BC' had already been noted by Flint and Deevey (1963). Curiously, this convention of subtracting 1950 years has persisted long after calibration scales have demonstrated that it is inappropriate.

DEVELOPMENTS IN RADIOCARBON SINCE THE NEED
FOR CALIBRATION WAS PERCEIVED

Since the 1960s various calibration curves and tables have been published, based on measurements of the radiocarbon concentration in tree-rings of precisely known calendar age (e.g. Suess 1970; Clark 1975; Klein *et al.* 1982). These calibrations seek to establish the relationship between the radiocarbon and calendrical timescales, using the annual tree-rings as the bridge. Most of the early published calibrations were based on results from one or more sets of measurements made on a single series of Bristlecone pine samples from California. The various treatments of the data yielded different curves depending on the smoothing functions used, but the general trend is the same in all of them. It is clear that the radiocarbon timescale is distorted in comparison with the linear calendar scale, and that the divergence varies with age.

In recent years much progress has been made in dendrochronological studies, particularly in the linking of the American Bristlecone pine and European oak chronologies. This has raised the possibility of new high precision calibration tables, which have begun to appear for some parts of the time scale (Stuiver 1982; Pearson and Baillie 1983. Final versions for the period 2000 BC to 1980 AD are in Kra and Stuiver 1986).

Once the need for calibration became accepted, the question arose as to how to refer unambiguously to calibrated and uncalibrated dates. The term BP already had a defined usage, universally accepted by radiocarbon scientists, and was therefore simply unavailable for an alternative use (in spite of developments mentioned below). The 1976 Radiocarbon conference at Los Angeles and San Diego resolved to label calibrated ages as AD*/BC* (Berger and Suess 1979: xi-xii), but this convention did not come into general use.

An alternative scheme for labelling calibrated and uncalibrated dates had been introduced by Suess and Strahm (1970), employing a suggestion by D. Shove. This paper relied on separate uses of ad/bc and AD/BC, but it is important to note that it did not make any alteration to the previously accepted usage of BP. An uncalibrated scale of the 'ad/bc' kind was attractive to archaeologists in Europe, because they were stuck with a large body of pre-calibration era literature, in which raw dates had been converted simply by converting 1950 (radiocarbon) years. They have also noted the competing claims of different calibration scales through the last fifteen years, and have needed an interim approach which harmonises at least approximately with the major historically based calendar. Nevertheless this usage has not gained approval among radiocarbon dating conventions (Stuiver and Polach 1977).

Subsequently, a parallel usage was introduced, in which 'bp' was taken to represent a raw radiocarbon age before 1950, and 'BP' used only for calibrated ages. This convention has come into use chiefly among archaeologists in Britain, and to a more limited extent in radiocarbon laboratories and scientific literature and radiocarbon laboratories closely involved with later period archaeology in Europe.

As a usage which 'evolved', it reflected the pressing need of archaeologists to know clearly which set of figures they were referring to. Unfortunately, this arrangement has engendered a number of difficulties:

(1) it redefines BP as accepted by radiocarbon daters on the basis of international conference decisions; their usage has scientific priority, and they have shown no inclination to change it;

(2)   more than half of radiocarbon users are in the earth sciences; they are less concerned with calibration than archaeologists, and have shown no move to alter the original BP usage;

(3)   in much of the World, including the Americas, where the Christian era is not a notable benchmark, archaeologists have held to the original BP usage;

(4)   the use of 'BP' to denote calibrated dates gives these an apparent respectability, though the great majority of dates have been calibrated on provisional curves which will soon be superseded.

The position has clearly been unsatisfactory, but there has been no one forum at which a consensus of opinion could easily emerge, either among archaeologists worldwide, or between archaeologists and radiocarbon-daters en masse. The position was considered carefully at the International Radiocarbon Conference at Trondheim in 1985, and the following conventions were agreed (Kra and Stuiver 1986):

(1)   BP is the exclusive designator for the timescale of radiocarbon years before 1950 (i.e. uncalibrated dates).
(2)   bp is not a recommended alternative.
(3)   Dates calibrated by an agreed curve can be referred to as 'Cal BP'.
(4)   Dates calibrated by an agreed curve can be converted to refer to the Christian era with the designator 'Cal AD' or 'Cal BC'.

THE TIMESCALES USED IN URANIUM SERIES, TL, K-AR AND FISSION TRACK

Other dating methods using different isotopes or different physical and chemical processes have different problems of terminology. None of those commonly used employs a defined zero year for reference and so their results cannot be directly related to the calendrical timescale. Methods such as thermoluminescence or uranium series are absolute in a theoretical sense, but this was also initially claimed for radiocarbon, and the need for recalibration of some datasets can never be ruled out.

Without some clear procedure for relating dates determined by these methods to the absolute calendrical scale, there is a potential problem with comparisons and concordance studies. Without a zero year there will eventually be a problem in relating dates determined decades apart. So far, the problems have not become serious, because most of the work has been carried out for periods where high precision is not required, or with relatively low precision. There clearly are different requirements for short and long timescales — a few decades are not crucial for dates of 10^4 to 10^6 years, but for more recent periods, better defined terms are certainly needed.

RECONCILING THE DIFFERENCES: CAN IT BE DONE?

Any system of referring to dates would ideally:

(a)   respect and be consistent with the needs of both the measurers and archaeological users;
(b)   allow consistency in relating timescales of different lengths;
(c)   be as consistent as possible with past published usage and existing conventions;
(d)   encourage 'good practice' (see Introduction).

The Trondheim recommendations should meet these needs for radiocarbon, and it is to be hoped that archaeologists will rapidly accept and use them, since there is little other chance of reaching consensus. ('Cal BC' and 'Cal AD' do of course indicate no more than estimates of 'real' AD/BC dates made with an excellent calibration curve: security of the estimate will depend upon the security of the date used.)

The principal need still not met is for an agreed universal scale of years, extending back from a fixed point. Evidently, AD/BC does not entirely meet this need, or the 'before present' terminologies would not have emerged. It would be natural to settle for 'BP' as a universal scale for all dating methods, but undeniably this convention is preempted by the raw radiocarbon scale. This is not just a theoretical difficulty: the Cambridge Earth Sciences 'Geologic time scale' (wrongly) takes for granted the wider applicability of 'BP' (Harland *et al.* 1982); thermoluminescence specialists will not be satisfied with the Trondheim recommendation that their dates should be expressed as 'years ago'; nor can they use the radiocarbon 'Cal BP', though there is no doubt that radiocarbon and TL dates will often need to be compared on the same diagrams.

It would be possible to use 'Cal BP' as 'Calendrical BP', but only if the 'Cal' in Cal AD and Cal BP above were discarded: Chris Chippindale is to make this suggestion on behalf of Antiquity (pers. comm.). We suggest therefore as an alternative 'ABP' for Absolute BP. It would be logical to retain AD 1950 as the zero year. One could use the SI term 'a', as in 'aBP'. This, however, would have disadvantages, as when '10 ka aBP' is converted to '10,000 a aBP', which is confusing. Thus we suggest that archaeologists and dating scientists might find a way forward by using an extra explanatory letter, with 'ABP' becoming a new and very plain universal scale.

In practice, of course, almost all dates except historical or dendrodates would be estimates projected onto this scale, rather than absolute values on the scale. This concept would avoid repeats of the problem encountered with radiocarbon, where 'BP' was assumed to be an absolute scale, and then shown not to be.

Now that radiocarbon daters have agreed to continue to use 'BP' for their raw scale, what will then happen to uncalibrated dates expressed in terms of the Christian epoch (the ad/bc usage)? The short answer is that this has been a stop-gap. We know that the radiocarbon timescale (BP) must be calibrated to achieve historically valid dates, and the approximate coincidence between 0 AD and 1950 BP is no more than fortuitous. It is therefore logically faulty to convert the raw BP scale to uncalibrated years relative to the Christian epoch (ad/bc) and the practice should be discouraged now that adequate calibrations are becoming available.

REFERENCES

Berger, R. and Suess, H.E. (eds.), 1979, Preface to *Radiocarbon dating, 9th Int. Radiocarbon Conf., Los Angeles and La Jolla, 1976, Proc.*, pp. xi-xii, Berkeley and Los Angeles: University of California Press.

Clark, R.M., 1975, A calibration curve for radiocarbon dates, *Antiquity* 49, 251–266.

De Vries, H., 1958, Variations in concentration of radiocarbon with time and location on earth, *Koninkl Ned Akad Wetensch Procd*, B61, 94–102.

Flint, R.F. and Deevey, E.S., 1961, Editorial Statement, *Radiocarbon* 3.

Flint, R.F. and Deevey, E.S., 1962, Editorial Statement, *Radiocarbon* 4.

Flint, R.F. and Deevey, E.S., 1963, Editorial Statement, *Radiocarbon* 5.

Harland, W.B., Cox, A.V., Llewellyn, P.G., Pickton, C.A.G., Smith, A.G. and Walters, R., 1982, *A geologic time scale*, Cambridge Earth Science Series, Cambridge: C.U.P.

Klein, J., Lerman, J.C., Damon, P.E. and Ralph, E.K., 1982, Calibration of radiocarbon dates: tables based on the consensus data of the Workshop on Calibrating the Radiocarbon Time Scale, *Radiocarbon* 24, 103–150.

Kra, R.R. and Stuiver, M., 1986, (eds.) *12th Int. Radiocarbon Conf., Trondheim, Norway, 1985, Proc., Radiocarbon* 28, 2.

Libby, W.F., 1952, 1955, *Radiocarbon Dating*, Chicago: University of Chicago Press.

Mann, W.B., Marlow, W.F. and Hughes, E.E., 1961, The half-life of carbon-14, *Jour. Applied Rad. Iso.* 11, 57–67.

Pearson, G.W. and Baillie, M.G.L., 1983, High-precision ^{14}C measurement of Irish oaks to show the natural atmospheric ^{14}C variations of the AD time period, in *11th Int. Radiocarbon Conf., Seattle, 1982, Proc.* (eds. Stuiver, M. and Kra, R.), *Radiocarbon 25*, 187–196.

Stuiver, M., 1982, A high-precision calibration of the AD radiocarbon time scale, *Radiocarbon* 24, 1–26.

Stuiver, M. and Polach, H.A., 1977, Discussion: reporting of ^{14}C data: *Radiocarbon* 19, 355–363.

Suess, H. and Strahm, C., 1970, The Neolithic of Auvernier: *Antiquity* 44, 91–95.

Zeuner, F.E., 1946, *Dating the past: an introduction to geochronology*, London: Methuen.

THE FUTURE PROSPECTS OF ACCELERATOR DATING

R. E. M. Hedges

Three major considerations come to mind that determine the future course of accelerator dating. These are: the resources available, technical developments, and the context in which the dates have their value. This paper discusses these in turn, from the position of managing the radiocarbon accelerator laboratory.

RESOURCES

Radiocarbon dating is a fairly labour-intensive process, particularly in the early stages of the sampling, documentation and pre-treatment of each individual sample. The measurement by accelerator requires, as is well known, the dedication of expensive equipment involving substantial maintenance, but capable of measuring a large number of samples per year. Present costs should come down by perhaps a factor of up to 2 if the maximum use of accelerater time is made. But the cost of a date will always be substantial (e.g. £200 at least), and the tendency of accelerator dating is to multiply the number of samples, since a more detailed dating strategy becomes possible. At present most of the cost of dating is provided by SERC, as part of its support for science-based archaeology. This support inevitably concentrates on developing the technology, and on dating projects which help to advance the methodology of radiocarbon dating. To support an increased turnover, and to ensure that archaeological programmes are chosen for their own merit, sources of funding from within the archaeological community will be required. This implies a certain unity of approach — something which the Conference is designed to foster.

TECHNICAL DEVELOPMENTS

The detection of ^{14}C by acceleration mass spectrometry (AMS) rather than by radioactivity measurements (conventional method) removes to a great extent some of the limits in conventional dating. At present the limits to dating by AMS are not those inherent in the method, but are set by the less than optimum operational procedures. These vary somewhat between AMS laboratories, but in general the present limits can be summarised:

Present performance limits error: ± 80 yrs (for last 10,000 yrs)
ultimate age: 30,000 yrs for useful dates; 40,000 yrs limit
sample size: 5 mg carbon optimal, 1 mg limit

Performance is also limited by the presence of contaminating material of different age in the sample. Although the small sample requirement of AMS dating is much better placed to deal effectively with such contamination, it remains an important problem, and one that can only be solved with a great deal of experiment and field experience.

As operational procedures improve, all these limits can be reduced. I doubt very much if the accuracy will ever equal that of 'high precision' conventional dating, however, although I think it is reasonable to expect ± 40 yrs (i.e. 1/2% error in the measurement). In most cases at this level other errors, to do with homogeneity of sample, and the calibration curve, become at least as important.

We know that the limitation on ultimate age at present comes from the sample processing chemistry. The ultimate limitation will be field contamination by younger carbon, but this will be highly variable, and at least in some cases may well permit dates beyond 60,000 years to be measured. However, because of the doubling of sensitivity to contamination every 6000 years, the battle will be a hard fight, and even 50,000 years seems a difficult goal at present.

The limitation on sample size is again set by the methods used in the sample chemistry, and there is no inherent reason why samples ten times smaller should not eventually be measurable. At whatever range of sample size one chooses, there will always be some samples too small for measurement.

The improvements indicated above can be expected generally; in the Oxford Laboratory we are developing an ion source to run from CO_2, rather than the more difficult to produce graphite which has been necessary in the past. Although too early to say yet, the CO_2 source should be able to improve on the three limitations mentioned above, as well as bring about an increase in the sample production rate, since the process of making CO_2 is so much simpler. But the proof of this lies in the future.

Chemical studies of the material dated, and the changes which it undergoes during burial are important in making much clearer exactly what has been dated, and in generally increasing the reliablity of associating the measured ^{14}C age with the event which the sample is supposed to represent. Improvements in reliability will not become dramatically evident, but should help to make the database of AMS dates a much more useful and long-lasting foundation for archaeological chronologies.

CHOOSING AND USING A RADIOCARBON DATE

The principal advantage of radiocarbon dating by AMS is that very much greater choice can be exercised over exactly what is being measured. This choice includes several different archaeological considerations such as degree of association, certainty of stratification, uniqueness, random sampling, and type of material, as well as more technical choices such as choice of chemical constituent, comparison of different 'fractions' and the ability to make replicate measurement on one object. How this choice is made depends on a keen appreciation of the strengths and limitations of radiocarbon dating.

Such an appreciation has several elements:

(1) Taking account of possible movement of the sample in relation to the site stratigraphy or other physical structure;

(2) Taking account of the different possible sources of carbon within the selected sample, and of the possible movement of carbon bearing constituents to and from the sample;

(3) Taking realistic account of the estimated error in the sample date (based on the ^{14}C measurement), including the reliability that the sample represents what is considered to be being dated, and combining the error given by the calibration curve;

(4) Establishing, or relating the date to, a chronological framework which is reasonably free from circular argument, and for which the errors have been realistically assessed.

All these points require the judgement of experience, although the information available for (3) allows it to enjoy the greatest objectivity. Clearly as more and more AMS dates are published, and experience grows, better understanding of the best sampling strategies and interpretation of results becomes possible. However, this will happen only if it is realised that AMS dating does bring about fresh approaches, and that in the past the points listed above have often not been adequately appreciated. It would be too bad if AMS dating was seen to be merely a technique available for those samples left over after conventional dating had been done.

The future development, then, depends in important ways on good, coherent and sympathetic communication between the laboratory and those selecting the samples and using the dates. Beyond that is the need to make a careful study of the complete dating process in action, and to learn how best to apply the lessons to radiocarbon dating in the future. This Conference is itself a very good example of one way of doing this.

APPENDIX: DATES PUBLISHED IN ARCHAEOMETRY

Further details of many dates cited in this book can be found in *Archaeometry* Datelists. This is a complete index to the first four datelists, arranged in order of laboratory numbers. The figures in brackets to the right refer to datelists as follows:

(1) = *Archaeometry* 26(1), 1984   (3) = *Archaeometry* 28(1), 1986

(2) = *Archaeometry* 27(2), 1985   (4) = *Archaeometry* 28(2), 1986

OxA-101	Wadi Kubbaniya	350 ± 200 (1)	OxA-147	Saqqara	4120 ± 150 (2)
OxA-102	Wadi Kubbaniya	101.5 ± 2.5% (1)	OxA-148	Pincevent	12600 ± 200 (3)
	(modern)		OxA-149	Pincevent	12400 ± 200 (3)
OxA-103	Wadi Kubbaniya	17150 ± 300 (1)	OxA-150	Poulton-le-Fylde	12400 ± 300 (2)
OxA-104	Guitarrero Cave	9930 ± 300 (3)	OxA-151	Poulton-le-Fylde	21500 ± 250 (2)
OxA-105	Monte Verde	12000 ± 250 (2)	OxA-152	Stanford man	4850 ± 150 (2)
OxA-106	Montecito	780 ± 150 (1)	OxA-153	Stanford man	4950 ± 150 (2)
OxA-107	Montecito	820 ± 150 (1)	OxA-154	San Diego Site W-12	8470 ± 140 (2)
OxA-108	Guitarrero Cave	10000 ± 200 (1,3)	OxA-157	Offham horse	2200 ± 120 (2)
OxA-109	Guitarrero Cave	9860 ± 200 (3)	OxA-173	Etiolles	12800 ± 220 (3,4)
OxA-110	Guitarrero Cave	2150 ± 150 (1,3)	OxA-174	Etiolles	11900 ± 250 (3)
OxA-111	Kendrick's Cave	10100 ± 200 (2)	OxA-175	Etiolles	12900 ± 220 (3)
OxA-112	Cairo Islamic doors	580 ± 150 (1)	OxA-176	Pincevent	12000 ± 220 (3)
OxA-113	Cairo Islamic doors	580 ± 150 (1)	OxA-177	Pincevent	12300 ± 220 (3)
OxA-114	Lindow Moss	1740 ± 100 (2)	OxA-178	Marsangy	11600 ± 200 (3)
OxA-115	Sinnock (modern)	100.0 ± 1.0% (1,2)	OxA-179	Montigny-sur-Loing	22200 ± 600 (3)
OxA-116	Sinnock	8030 ± 160 (1,2)	OxA-180	Montigny-sur-Loing	22500 ± 600 (3)
OxA-117	Vaenge	5475 ± 130 (1)	OxA-181	Guitarrero Cave	9520 ± 150 (3)
OxA-118	Holmegaard-Jutland	6200 ± 130 (1)	OxA-182	Guitarrero Cave	9280 ± 150 (3)
OxA-119	Dorchester bypass	4800 ± 130 (1)	OxA-183	Guitarrero Cave	9340 ± 150 (3)
OxA-120	Cyprus slag	2860 ± 150 (1)	OxA-184	Guitarrero Cave	9400 ± 150 (3)
OxA-121	Jackscar Cave	2670 ± 120 (2)	OxA-185	Guitarrero Cave	9350 ± 150 (3)
OxA-127	BM Comparison No.1	4390 ± 100 (3)	OxA-186	La Jolla	5600 ± 400 (2)
OxA-128	BM Comparison No.2	300 ± 100 (3)	OxA-187	Sunnyvale	6350 ± 400 (2)
OxA-129	Pompeii	2000 ± 130 (2,3)	OxA-188	Del Mar	5400 ± 120 (2)
OxA-133	BM Comparison No.3	630 ± 80 (3)	OxA-189	Laguna	5100 ± 500 (2)
OxA-134	Pompeii	2020 ± 160 (2,3)	OxA-191	Epirus, Voidomatis	1000 ± 150 (2)
OxA-135	Shiqmim	210 ± 150 (2)	OxA-192	Epirus, Voidomatis	800 ± 100 (2)
OxA-136	Klithi Rockshelter	16300 ± 400 (2)	OxA-193	Guitarrero Cave	9600 ± 130 (3)
OxA-137	Klithi Rockshelter	17000 ± 400 (2)	OxA-194	Guitarrero Cave	9430 ± 150 (3)
OxA-138	Etiolles	12990 ± 300 (3)	OxA-195	Guitarrero Cave	10180 ± 130 (3)
OxA-139	Etiolles	13000 ± 300 (3)	OxA-196	Guitarrero Cave	9980 ± 120 (3)
OxA-140	Paviland	38800 ± 8000 (3)	OxA-197	Guitarrero Cave	10340 ± 130 (3)
OxA-141	Weelde-Paardsdrank	8160 ± 150 (2)	OxA-198	Guitarrero Cave	100.0 ± 1.2% (3)
OxA-142	Weelde-Paardsdrank	7090 ± 150 (2)		(modern)	
OxA-143	Weelde-Paardsdrank	3330 ± 130 (2)	OxA-199	Robin Hood's Cave	> 36000 (3)
OxA-144	Danebury	2180 ± 130 (2)	OxA-351	Epirus, Voidomatis	11100 ± 200 (2)
OxA-145	Gordion	2650 ± 150 (2)	OxA-352	Epirus, Voidomatis	15840 ± 250 (2)
OxA-146	Mycenae	2970 ± 130 (2)	OxA-353	Epirus, Voidomatis	10700 ± 200 (2)

166

OxA-361	Cuello	2460 ± 80 (3,4)
OxA-362	Cuello	2390 ± 90 (3,4)
OxA-363	Meadowcroft	31400 ± 1200 (2)
OxA-364	Meadowcroft	30900 ± 1100 (2)
OxA-365	Paviland	29600 ± 1900 (3)
OxA-366	Paviland	28000 ± 1700 (3)
OxA-371	Offham horse	2280 ± 120 (2)
OxA-372	Epirus, Voidomatis	14600 ± 340 (2)
OxA-375	Jebel Naja	7430 ± 100 (4)
OxA-376	Longmoor	8930 ± 100 (2)
OxA-377	Longmoor	8760 ± 110 (2)
OxA-378	Kettlebury	8270 ± 120 (2)
OxA-379	Kettlebury	7940 ± 120 (2)
OxA-380	Robin Hood's Cave	4250 ± 75 (3)
OxA-381	Monte Verde	12400 ± 150 (2)
OxA-382	Robin Hood's Cave	3100 ± 80 (3)
OxA-383	Hazleton	4450 ± 90 (2)
OxA-385	Lac du Bonnet	920 ± 100 (2)
OxA-388	Can Hasan	7910 ± 160 (2)
OxA-391	Pincevent	11870 ± 130 (3)
OxA-392	Can Hasan	250 ± 90 (2)
OxA-393	Wadi Hammeh	11920 ± 150 (4)
OxA-394	Wadi Hammeh	12200 ± 160 (4)
OxA-398	Hengistbury	8590 ± 120 (2)
OxA-399	Hengistbury	4770 ± 180 (2)
OxA-401	La Ferrassie	23800 ± 530 (4)
OxA-402	La Ferrassie	27900 ± 770 (4)
OxA-403	La Ferrassie	27530 ± 720 (4)
OxA-404	La Ferrassie	26250 ± 620 (4)
OxA-405	La Ferrassie	29000 ± 850 (4)
OxA-409	La Ferrassie	28600 ± 1050 (4)
OxA-410	Combe Saunière	15750 ± 230 (4)
OxA-411	Hengistbury	7690 ± 110 (2)
OxA-412	Hengistbury	8140 ± 120 (2)
OxA-413	Hengistbury	7910 ± 140 (2)
OxA-414	Pincevent IV	8850 ± 130 (3)
OxA-416	Peak Camp	4630 ± 110 (2)
OxA-417	Peak Camp	4660 ± 80 (2)
OxA-418	Cairo MS	1040 ± 60 (2)
OxA-419	Nablus MS	970 ± 80 (2)
OxA-420	Amalfi MS	750 ± 50 (2)
OxA-421	Mappa Mundi	850 ± 60 (2)
OxA-424	Mary Rose	300 ± 60 (2)
OxA-425	Pompeii	1960 ± 80 (2)
OxA-444	Peak Camp	4790 ± 80 (2)
OxA-445	Peak Camp	4670 ± 90 (2)
OxA-446	Peak Camp	4810 ± 90 (2)
OxA-447	Le Flageolet	25700 ± 700 (4)
OxA-448	Le Flageolet	24600 ± 700 (4)
OxA-449	West Kennet	4825 ± 90 (2)
OxA-450	West Kennet	4700 ± 80 (2)
OxA-451	West Kennet	4780 ± 90 (2)
OxA-459	Combe Saunière	15480 ± 210 (4)
OxA-460	Belloy-sur-Somme	5255 ± 5255 (4)
OxA-461	Belloy-sur-Somme	8010 ± 110 (4)
OxA-462	Belloy-sur-Somme	9720 ± 130 (4)
OxA-463	Gough's New Cave	12380 ± 160 (2)
OxA-464	Gough's New Cave	12470 ± 160 (2)
OxA-465	Gough's New Cave	12360 ± 170 (2)
OxA-466	Gough's New Cave	12800 ± 170 (2)
OxA-467	Pincevent	12250 ± 160 (3)
OxA-480	Laugerie-Haute Est	14730 ± 250 (4)
OxA-481	Combe Saunière	14990 ± 220 (4)
OxA-482	Combe Saunière	26920 ± 800 (4)
OxA-485	Combe Saunière	16300 ± 220 (4)
OxA-486	Combe Saunière	22100 ± 440 (4)
OxA-487	Combe Saunière	10140 ± 120 (4)
OxA-488	Combe Saunière	17700 ± 290 (4)
OxA-489	Combe Saunière	19450 ± 330 (4)
OxA-490	Pincevent	7140 ± 100 (3)
OxA-491	Pincevent	3650 ± 100 (3)
OxA-492	Laugerie-Haute Est	14770 ± 180 (4)
OxA-493	Son Muleta Cave	36000 ± 2500 (4)
OxA-494	Son Muleta Cave	10500 ± 500 (4)
OxA-495	Son Muleta Cave	14150 ± 200 (4)
OxA-496	Son Muleta Cave	12890 ± 180 (4)
OxA-497	Son Muleta Cave	12020 ± 160 (4)
OxA-498	Son Muleta Cave	9000 ± 3000 (4)
OxA-499	Son Muleta Cave	12100 ± 180 (4)
OxA-500	Northampton harpoon	¯9240 ± 160 (3)
OxA-501	Klithi Rockshelter	260 ± 100 (3)
OxA-502	Klithi Rockshelter	12300 ± 200 (3)
OxA-503	Pompeii	2000 ± 80 (3)
OxA-504	Orton Gravel Complex	3390 ± 100 (4)
OxA-505	Marsangy	9770 ± 180 (3,4)
OxA-506	Tree ring AD 1000	980 ± 60 (3)
OxA-507	Wadi Hammeh	11950 ± 160 (4)
OxA-509	Kariya Wuro	220 ± 50 (3)
OxA-510	Uchcumachay	6670 ± 140 (3)
OxA-511	Uchcumachay	890 ± 90 (3)
OxA-512	Lake Gramousti	7270 ± 120 (3)
OxA-513	Rezina Marsh	5000 ± 100 (3)
OxA-514	Mary Rose	290 ± 70 (3)
OxA-516	Great Doward pika	10020 ± 120 (3)
OxA-517	Sproughton Point 1	10910 ± 150 (3)
OxA-518	Sproughton Point 2	10700 ± 160 (3)
OxA-519	Wadi el Jilat	21150 ± 400 (4)
OxA-520	Wadi el Jilat	14790 ± 200 (4)
OxA-521	Wadi el Jilat	13310 ± 120 (4)
OxA-522	Wadi el Jilat	11740 ± 80 (4)
OxA-523	Wadi el Jilat	11450 ± 200 (4)
OxA-524	Wadi el Jilat	15520 ± 200 (4)
OxA-525	Wadi el Jilat	16010 ± 200 (4)
OxA-526	Wadi el Jilat	8810 ± 110 (4)
OxA-527	Wadi el Jilat	8520 ± 110 (4)
OxA-528	Tree ring AD 1000	890 ± 60 (3)
OxA-529	Pompeii	1950 ± 70 (3)

OxA-535	Sun Hole	12210 ± 160 (3)
OxA-538	Grotte de Bange	12080 ± 180 (3)
OxA-539	Wadi el Jilat	7980 ± 150 (4)
OxA-540	Grotte de Bange	12200 ± 160 (3)
OxA-542	Klithi Rockshelter	10420 ± 150 (3)
OxA-546	Tree ring AD 1000	1030 ± 50 (3)
OxA-563	West Kennet	4780 ± 90 (4)
OxA-567	Newlands Cross	7600 ± 900 (4)
OxA-568	Newlands Cross	7600 ± 500 (4)
OxA-569	Newlands Cross	9720 ± 300 (4)
OxA-570	Kariya Wuro	960 ± 300 (3)
OxA-578	Mukutan River	2600 ± 80 (3)
OxA-579	Le Flageolet	26500 ± 900 (4)
OxA-583	Abri du Facteur	24720 ± 600 (4)
OxA-584	Abri du Facteur	24210 ± 500 (4)
OxA-585	Abri du Facteur	24400 ± 600 (4)
OxA-586	Abri du Facteur	24690 ± 600 (4)
OxA-587	Gough's Old Cave	12530 ± 150 (3)
OxA-588	Gough's New Cave	12030 ± 150 (3)
OxA-589	Gough's New Cave	12340 ± 150 (3)
OxA-590	Gough's New Cave	12370 ± 150 (3)
OxA-591	Gough's New Cave	12260 ± 160 (3)
OxA-592	Gough's New Cave	12500 ± 160 (3)
OxA-593	Broxbourne	7230 ± 150 (3)
OxA-594	Abri du Facteur	25450 ± 650 (4)
OxA-595	Abri du Facteur	25630 ± 650 (4)
OxA-596	Le Flageolet	23250 ± 500 (4)
OxA-597	Le Flageolet	24800 ± 600 (4)
OxA-598	Le Flageolet	33800 ± 1800 (4)
OxA-601	Misbourne	6190 ± 90 (4)
OxA-602	Misbourne	3730 ± 90 (4)
OxA-603	Misbourne	4070 ± 100 (4)
OxA-606	Pompeii	1930 ± 80 (3)
OxA-613	Qasr Ibrim	570 ± 200 (3)
OxA-614	Tree ring AD 1000	1000 ± 80 (3)
OxA-618	Misbourne	5970 ± 100 (4)
OxA-619	Misbourne	6100 ± 120 (4)
OxA-620	Misbourne	2500 ± 150 (4)
OxA-621	Misbourne	12530 ± 200 (4)
OxA-624	Dorset Cursus	4570 ± 120 (3)
OxA-625	Dorset Cursus	4440 ± 100 (3)
OxA-626	Dorset Cursus	4770 ± 120 (3)
OxA-627	Dorset Cursus	6800 ± 100 (3)
OxA-628	Dorset Cursus	6460 ± 140 (3)
OxA-629	Ossom's Cave	2030 ± 80 (3)
OxA-630	Ossom's Cave	4860 ± 80 (3)
OxA-631	Ossom's Cave	10780 ± 160 (3)
OxA-632	Ossom's Cave	10600 ± 140 (3)
OxA-635	Krakow	20200 ± 350 (4)
OxA-636	Wadi el Jilat	10540 ± 160 (4)
OxA-638	Peak Camp	4290 ± 80 (4)

OxA-643	Hazleton	4600 ± 120 (3)
OxA-644	Hazleton	4840 ± 80 (3)
OxA-645	Hazleton	4780 ± 80 (3)
OxA-646	Hazleton	4875 ± 80 (3)
OxA-659	Tree ring AD 1000	1020 ± 60 (3)
OxA-660	Great Domesday	990 ± 60 (4)
OxA-661	Great Domesday	550 ± 65 (4)
OxA-662	Great Domesday	300 ± 80 (4)
OxA-663	Great Domesday	220 ± 70 (4)
OxA-664	Great Domesday	1160 ± 100 (4)
OxA-665	Little Domesday	1140 ± 70 (4)
OxA-666	Little Domesday	660 ± 60 (4)
OxA-667	Little Domesday	480 ± 60 (4)
OxA-677	Tree ring AD 1000	980 ± 60 (3)
OxA-679	Badger Hole	9060 ± 130 (4)
OxA-680	Badger Hole	1380 ± 70 (4)
OxA-681	Paviland	7190 ± 80 (4)
OxA-691	Soldier's Hole	> 34500 (4)
OxA-692	Soldier's Hole	29300 ± 1100 (4)
OxA-693	Soldier's Hole	> 35000 (4)
OxA-694	Soldier's Hole	19300 ± 400 (4)
OxA-707	Newlands Cross	8300 ± 90 (4)
OxA-708	Newlands Cross	8930 ± 150 (4)
OxA-722	Belloy-sur-Somme	10110 ± 130 (4)
OxA-723	Belloy-sur-Somme	9890 ± 150 (4)
OxA-724	Belloy-sur-Somme	10260 ± 160 (4)
OxA-730	Moque Panier	12300 ± 160 (4)
OxA-731	Moque Panier	12240 ± 160 (4)
OxA-735	Church Hole Cave	12240 ± 150 (4)
OxA-736	Robin Hood's Cave	2020 ± 80 (4)
OxA-740	Le Pré des Forges	12120 ± 200 (4)
OxA-742	Hayonim	12360 ± 160 (4)
OxA-743	Hayonim	12010 ± 180 (4)
OxA-744	Netiv Hagdud	9700 ± 150 (4)
OxA-747	Klithi	3560 ± 1000 (4)
OxA-748	Klithi (modern)	101.8 ± 1.0% (4)
OxA-749	Klithi	14200 ± 200 (4)
OxA-750	Klithi	14060 ± 200 (4)
OxA-751	Combe Saunière	15190 ± 200 (4)
OxA-752	Combe Saunière	19490 ± 350 (4)
OxA-753	Combe Saunière	19630 ± 320 (4)
OxA-754	Combe Saunière	15200 ± 200 (4)
OxA-755	Combe Saunière	14890 ± 200 (4)
OxA-756	Combe Saunière	15120 ± 200 (4)
OxA-757	Combe Saunière	18860 ± 320 (4)
OxA-758	Combe Saunière	21640 + 400 (4)
OxA-759	Laugerie-Haute Est	14320 ± 180 (4)
OxA-760	Laugerie-Haute Est	15730 ± 200 (4)
OxA-761	Laugerie Haute-Est	14320 ± 180 (4)
OxA-762	Laugerie Haute-Est	14100 ± 100 (4)
OxA-763	Tree ring AD 1887	90 ± 60 (4)

OxA-764	Tree ring AD 3995	5220 ± 80 (4)
OxA-766	Tree ring AD 490	1520 ± 100 (4)
OxA-767	Tree ring AD 1887	120 ± 70 (4)
OxA-768	Combe Saunière	14260 ± 200 (4)
OxA-769	Combe Saunière	14800 ± 240 (4)
OxA-770	Combe Saunière	14770 ± 200 (4)
OxA-773	Taber infant	3390 ± 90 (4)
OxA-774	Del Mar	5720 ± 100 (4)
OxA-779	Tree ring AD 10	1910 ± 80 (4)
OxA-780	Tree ring AD 490	1920 ± 100 (4)
OxA-792	Pompeii	2000 ± 100 (4)
OxA-793	Mary Rose	360 ± 80 (4)
OxA-795	Tree ring AD 10	2040 ± 10 (4)
OxA-796	Tree ring 995 BC	2840 ± 80 (4)
OxA-799	Aveline's Hole	9100 ± 100 (4)
OxA-800	Aveline's Hole	8860 ± 100 (4)
OxA-801	Aveline's Hole	12100 ± 180 (4)
OxA-802	Aveline's Hole	9670 ± 110 (4)
OxA-803	Lyngby axe	10320 ± 150 (4)
OxA-811	Elder Bush Cave	10600 ± 110 (4)
OxA-812	Elder Bush Cave	9000 ± 130 (4)
OxA-813	Gough's New Cave	11900 ± 140 (4)
OxA-814	Gough's New Cave	9100 ± 100 (4)
OxA-815	Gough's New Cave	1740 ± 60 (4)
OxA-820	Little Domesday	420 ± 90 (4)
OxA-825	Mary Rose	310 ± 80 (4)
OxA-826	Tree ring AD 1000	1050 ± 80 (4)
OxA-874	Tree ring AD 485	1600 ± 90 (4)
OxA-875	Tree ring 3995 BC	5140 ± 100 (4)
OxA-901	Tree ring AD 1000	2810 ± 60 (4)
OxA-917	Temple of Tuthmosis IV	3150 ± 80 (4)

CONTRIBUTORS

S.H. Andersen
Institut for forhistorisk arkaeologi, Aarhus
Universitet, Moesgård, DK-8270 Højbjerg,
Denmark.

G.N. Bailey
Department of Archaeology, University of
Cambridge, Downing Street, Cambridge.

R.N.E. Barton
Donald Baden-Powell Quaternary Research Centre,
60 Banbury Road, Oxford OX2 6PN.

R. Bradley
Department of Archaeology, University of
Reading, Whiteknights, Reading RG6 2AA.

H.M. Bricker
Department of Anthropology, Tulane University,
New Orleans, Louisiana 70118, U.S.A.

R. Burleigh
Museum of Mankind, The British Museum,
Burlington Gardens, London W1X 2EX.

T.S. Constandse-Westermann
Institute of Human Biology, Achter de Dom 34,
Utrecht, Netherlands.

J. Cook
Department of Prehistoric and Romano-British
Antiquities, The British Museum, London WC1.

T.C. Darvill
209 Seymour Road, Gloucester GL1 5HR.

J.G. Evans
Department of Archaeology, University College,
P.O. Box 78, Cardiff CF1 1XL.

C.S. Gamble
Department of Archaeology, University of
Southampton, Southampton, SO9 5NH.

R. Gillespie
33 Charlotte Street, Leichhardt, NSW 2040,
Australia.

J.A.J. Gowlett
Radiocarbon Accelerator Unit, Research
Laboratory for Archaeology, University of Oxford,
6 Keble Road, Oxford OX1 3QJ.

D.R. Harris
Department of Human Environment, Institute of
Archaeology, University of London, 31–34 Gordon
Square, London WC1H OPY.

R.E.M. Hedges
Radiocarbon Accelerator Unit, Research
Laboratory for Archaeology, University of Oxford,
6 Keble Road, Oxford OX1 3QJ.

H.P. Higgs
Department of Archaeology, University of
Cambridge, Downing Street, Cambridge.

R.M. Jacobi
Department of Classics & Archaeology, University
of Lancaster, Bailrigg, Lancaster LA1 4YN.

A.J. Legge
Department of Extra-Mural Studies, University of
London, 26 Russell Square, London WC1B 5DQ.

P.A. Mellars
Department of Archaeology, University of
Cambridge, Downing Street, Cambridge CB2 3DZ.

S.P. Needham
Department of Prehistoric and Romano-British
Antiquities, The British Museum, London WC1.

R.R. Newell
Biologisch-Archaeologisch Instituut,
Rijksuniversiteit, 9712 ER Groningen, Poststraat 6,
Netherlands.

C. Roubet
Institut de Paléontologie Humaine, 1 Rue René
Panhard, 75013 Paris, France.

P. Rowley-Conwy
Department of Extra-Mural Studies, University of
London, 26 Russell Square, London WC1B 5DQ.

A. Saville
Art Gallery and Museum, Clarence Street,
Cheltenham, Gloucestershire GL50 3NX.

D.D.A Simpson
Department of Archaeology, The Queen's
University, Belfast BT7 1PA, Northern Ireland.

O. Soffer
Department of Anthropology, University of
Illinois, 109 Davenport Hall, 607 South Mathews
Avenue, Urbana, Illinois 61801, U.S.A.

C.B. Stringer
Department of Palaeontology, British Museum
(Natural History), Cromwell Road, London SW7.

D.A. Sturdy
Bermuda Maritime Museum, P.O. Box 273,
Somerset 9, Bermuda.

C. Turner
Department of Earth Sciences, The Open
University, Walton Hall, Milton Keynes.

D.P. Webley
Department of Archaeology, University College,
P.O. Box 78, Cardiff CF1 1XL.